"The time is right for an insightful, well-documented exposé of the pathology in poverty neighborhoods and a road map for the journey toward health and wholeness. *How Neighborhoods Make Us Sick* is just that. Turning troubled communities around is no small challenge, the authors admit. But there are practical steps that have proven to be effective. This is essential reading for anyone engaged in service among the poor."

Bob Lupton, community developer, author of *Toxic Charity*

"Framed expertly in terms of the macrolevel social determinants of health and the gap in life expectancy between poor neighborhoods and wealthier ones, this heartfelt first-person account by two staff members of Atlanta's Good Samaritan Health Center makes vivid the microlevel daily pain and struggles of those who live in poverty. It also outlines an activist, social justice approach to making the changes that have to be made with and by community members in order for neighborhoods to produce health, and not harm, to their residents."

Ellen Idler, director of religion and public health collaborative, Emory University

"In my thirty years of working in health nonprofits, I've never come across a book that puts a real face on the socioeconomic and geographic disparities that truly exist in health care. I recommend this book to anyone interested in the stark realities of providing health services to vulnerable populations from two compassionate, remarkable women working and living in the trenches, while providing a roadmap to community wellness."

Donna Looper, executive director of Georgia Charitable Care Network Inc.

"At a time where words like poverty, injustice, mental health issues, trauma, and the like have become familiar to our common justice narrative, we are in need of a deeper dive into how these systemic barriers truly impact communities. Veronica and Breanna have generously invited us into their realm by sharing their experiences and learned expertise in an effort to help illuminate the need for deeper awareness and collective action toward the flourishing of poor communities."

Michelle Ferrigno Warren, advocacy and strategic engagement director of the CCDA, author of *The Power of Proximity*

"Through compelling stories, this eye-opening book illuminates the inverse relationship between poverty and health of all kinds. It acknowledges both the complexities and hard realities in addressing needs that often feel overwhelming. And ultimately, it points to hopeful strategies that flow from taking ownership of both the problem and the solution."

Amy Simpson, author of *Troubled Minds*

"*How Neighborhoods Make Us Sick* was a disturbing read because in its pages I saw descriptions of where my two young-adult, adopted, biracial sons potentially would be if they did not have the safety net of financially secure parents behind them. The authors are realistic about the complexity of the problems that those with limited resources face. They avoid simplistic answers while offering possible solutions for greater overall health. Maybe there is hope after all."

Heather Davediuk Gingrich, professor of counseling at Denver Seminary, coeditor of *Treating Trauma in Christian Counseling*

"*How Neighborhoods Make Us Sick* is an excellent and desperately needed addition to the nursing literature on community and social justice. Impressive for its narrative style that provides accessibility, relevance, and a guide for praxis, Veronica Squires and Breanna Lathrop detail their journeys and desires to help the most vulnerable in society and offer us a glimpse into the transformation of mere practice into praxis—practice that is intentionally aimed at social justice goals. Their pathways to praxis are substantiated by their poignant and pragmatic accounts of emerging awareness, deep self-reflection, risk taking, choosing change, and taking action to transcend present circumstances for both their clients and themselves. A mandatory book for current times as nurses expand their scope and locus of practice into the community. *How Neighborhoods Make Us Sick* is an amazing example of emancipatory nursing. It will transfix and engage any health care provider seeking to reduce suffering by uncovering and challenging systemic barriers to better society and quality of life through true praxis."

Paula N. Kagan, associate professor at DePaul University, lead author of *Philosophies and Practices of Emancipatory Nursing*

VERONICA SQUIRES AND
BREANNA LATHROP

HOW NEIGHBORHOODS MAKE US SICK

Restoring Health and Wellness to Our Communities

IVP Books
An imprint of InterVarsity Press
Downers Grove, Illinois

InterVarsity Press
P.O. Box 1400, Downers Grove, IL 60515-1426
ivpress.com
email@ivpress.com

InterVarsity Press® is the book-publishing division of InterVarsity Christian Fellowship/USA®, a movement of students and faculty active on campus at hundreds of universities, colleges, and schools of nursing in the United States of America, and a member movement of the International Fellowship of Evangelical Students. For information about local and regional activities, visit intervarsity.org.

While any stories in this book are true, some names and identifying information may have been changed to protect the privacy of individuals.

Cover design: David Fassett
Interior design: Daniel van Loon
Images: apartment block: © duncan1890 / E+ / Getty Images
shot of city map: © Ugo Leonetti / /EyeEm /Getty Images

ISBN 978-0-8308-4557-6 (print)
ISBN 978-0-8308-7335-7 (digital)

Printed in the United States of America ♾

InterVarsity Press is committed to ecological stewardship and to the conservation of natural resources in all our operations. This book was printed using sustainably sourced paper.

Library of Congress Cataloging-in-Publication Data
A catalog record for this book is available from the Library of Congress.

P 23 22 21 20 19 18 17 16 15 14 13 12 11 10 9 8 7 6 5 4 3 2 1
Y 38 37 36 35 34 33 32 31 30 29 28 27 26 25 24 23 22 21 20 19

With admiration for our colleagues and contributors

who are making their neighborhoods healthier.

And for our children:

May our life work help narrow the

life expectancy gap for your generation.

CONTENTS

FOREWORD

KERI NORRIS, PhD, MPH, MCHES

In the race toward health equity, we have embraced many models in an effort to close gaps in health disparities for underserved and indigent populations. We know there is no one-size-fits-all solution. I submit that we must keep working toward health equity and embrace models that empower the communities we serve and move toward an all-inclusive policy and integrated care resolutions.

I personally believe that health equity is achieved when all levels of the socioecological model are addressed, but more specifically, the outer realms. We have spent so much time trying to change behavior that we forget once the behavior changes it is met with resistance at the institutional and policy levels. How can one be expected to maintain a positive change when met with obstacles at various levels? Once we embrace having everyone at the table—communities, housing, safety, urban planning, faith-based organizations, health systems, workforce development, legislators, and so forth, then we will move the needle on health equity. Collaborative partnerships are the most important tool along with policy change. That is how we win at health equity!

The real and lived experiences of Veronica and Breanna are indicative of how where you live impacts your mental and physical health, regardless of your socioeconomic status. They are dedicated servant leaders who have a heart toward health equity and made it their purpose to immerse themselves in communities with the greatest needs. I urge readers to forget the race of these women and do not liken them to well-to-do women who volunteer in disparate communities and leave it at donations and daytime exchanges with poor folks. This was no ethnographic study; this was the heart of Christ in two professional women who embedded themselves in walking the walk. Two professional women who now understand the communities they serve more than ever and who advocate for change even when it means challenging the status quo.

As a public health professional with over sixteen years of experience working in disparate and vulnerable communities throughout the United States and its territories, I compel you to see how models of health equity should begin to expand and encompass cultural humility and servant leadership and revisit community-based participatory programs. We can look to foundational initiatives, like the East Side Village Health Workers Partnership, and the local efforts highlighted in this book, such as the Good Samaritan Health Center and King County's Office of Equity and Social Justice, to find models that need replicating throughout our communities and will finally bridge the gap.

PART ONE

HOW WE GET SICK

CHAPTER 1

TWO JOURNEYS TO THE INNER CITY

But if I could turn life on its side
Go back and do everything right
Oh I think, I think
that I might.

JENN BOSTIC, "SNOWSTORM"

The drive to the clinic in the morning is usually dark, the sun just starting to emerge over the heart of West Atlanta. Near the clinic, the roads become lined with abandoned houses and closed businesses. There are several corner shops and a few gas stations. Despite the early hours, there is always foot traffic and kids in uniform waiting at bus stops. Occasionally stray dogs wander along the road, and someone holding a sign is asking for money. Shortly before the clinic opens, the local jail procession drives past taking prisoners to work duty for the day. The landscape fits the story of urban poverty: unemployment, crime, and abandonment.

Yet, having worked and lived in the neighborhood for nearly a decade, we are aware of another story as well. The neighborhood is filled with vibrant church communities offering fellowship, meals, and social services in addition to weekly sermons. There is an urban garden movement with fresh produce hiding behind many rundown buildings. Several clinics, like the one we work in, offer health care services and health education. Progressive and creative learning communities like BEST Academy and KIPP are just blocks off the route to work. The neighborhood is filled with engaged community members with deep passion for their community and strong family ties. Yet the neighborhood is suffering.

When the Virginia Commonwealth University released their Mapping Life Expectancy project, we saw numbers that confirmed what years of clinical and community development experience had revealed. Life expectancy in this West Atlanta zip code is thirteen years less than life expectancy on the more affluent side of town.[1] A twenty-minute drive within the perimeter of Atlanta creates a life expectancy gap of thirteen years. This finding wasn't unique to Atlanta. From a sixteen year gap in Chicago to twelve years in St. Louis and twenty years in Philadelphia, life expectancy gaps persist throughout US urban areas. How is it possible that people living within the same city could have such drastic differences in life expectancies? How can residents in neighboring zip codes have such distinct differences in mortality?

Similar questions have plagued us since starting our careers. Breanna is a family nurse practitioner who has worked in nonprofit clinics since graduating in 2008. In order to better understand the health care system, she obtained a master's degree in public health and focused her doctoral study on social determinants of health and programs to improve clinical care for vulnerable populations. In 2010, Breanna started working at the Good Samaritan Health Center (Good Sam), a nonprofit clinic providing comprehensive care to people without health insurance or adequate financial resources to afford health care. The clinic, now in its twentieth year, offers primary care, dental services, health education, behavioral health, and some specialty care in West Atlanta with a mission of "spreading Christ's love through quality health care to those in need." Veronica joined the Good Sam team in 2015. Veronica's career prior had focused on nonprofit fundraising and development. As a Christian Community Development practitioner, she lived in Southwest Atlanta years before working in a similar neighborhood.

We found that despite our career differences, we shared a deep concern for the health of the neighborhood and questioned whether our work, personally and professionally, was making any difference. Breanna had spent the first few years of her career learning the limitations of clinical medicine in restoring health to underresourced communities. She had met countless motivated patients who sacrificed to afford their medications and make it to appointments on foot in the rain because they had no other transportation. Yet even with their motivation and the best medical care she could provide, she saw the direct health impact of poverty and faced the sobering reality that health care simply wasn't enough. Veronica moved to Southwest Atlanta with her family in hopes of improving her community through youth development, civic engagement, and fundraising for capital expansion projects. Yet she and her family found the neighborhood was making them sick far faster than their efforts were improving the neighborhood.

Over our years of working together, we have had many long conversations about what is making the neighborhood sick. We have watched in frustration as patients face insurmountable barriers to good health and have recognized our limitations in dismantling such barriers. This book is our response to the question, What is making our neighborhood sick? It is also a discussion of the programs, policies, and community efforts that are bringing wellness and confronting the systems of oppression that allow such life expectancy gaps to exist. We also hope this book will be a catalyst for change and offer practical applications in part two for how you, the reader, can make a difference.

In the chapters that follow, we use our personal experiences along with the shared experiences of friends, neighbors, colleagues, and community partners to illustrate the ways social determinants affect health. We recognize that because we are white women from a middle-class background, we can never fully understand the challenges our neighbors and patients have experienced. We cannot identify with the complex layers of historical discrimination and generational poverty faced by many within our communities. Yet despite these limitations, we chose to write this book to illustrate the power and pervasiveness of social determinants such as poverty, homelessness, environment, and education to impact health status. The currently available literature on social determinants is often academic. We come with personal, real-life language to describe what these determinants are and how they can negatively impact health, disproportionately so in low-income communities. Our aim is not to point a finger and judge these communities for being sick. Rather, we are affirming the very real obstacles that exist for these residents and calling for large-scale change to heal their neighborhoods. We hope this is evident throughout the remainder of the book. We have tried to keep an attitude of humility, listening to and learning from our neighbors and patients about interventions most helpful to them. We also recognize that while this book focuses on experiences and issues in urban Atlanta, poverty, like a disease, spares no geographical region or race. Social determinants of health impact rural areas in different yet equally important ways.

A few years back a childhood friend of Breanna's was coming to Atlanta for work. In route to Breanna's house, he called her to say he was lost. "I think my GPS is trying to kill me!" he explained. In an attempt to reroute him, she quickly discovered he was just a block from the Good Sam clinic. His sentiment reflected the culture shock of driving through urban poverty, but the statement holds an element of truth. The neighborhood is deadly, but her lost friend passing through isn't the one who is at risk. For those living within the neighborhood, poverty, unemployment, racism, the built environment, and systems of oppression are literally making them sick.

VERONICA'S JOURNEY

I first heard about community redevelopment during my freshman year at Emory University a few months into joining the InterVarsity Christian Fellowship chapter, a diverse witnessing community of believers on campus. Having grown up in a small suburb outside of Orlando, Florida, going away to school at Emory in Atlanta was my first introduction to a big urban city. I fell in love with both.

As I became more involved in InterVarsity, I was encouraged to see my life as having a mission and that I could be a "world changer." I jumped in with both feet. It was like I had finally been acquainted with the missing link that made everything else in my life make sense—my helping personality, my love for people, and my desire to see broken systems restored to new life. Surrounding me were Christians acting as Jesus' hands and feet by caring about racial reconciliation, loving the poor, engaging with those who mourn, and fighting for justice. I came alive as a Christian and started devouring books by C. S. Lewis, Dietrich Bonhoeffer, Dorothy Day, and Elisabeth Elliot. My list of spiritual heroes grew tremendously in college.

It helped that in the larger InterVarsity network there were many examples of individuals and families making radical decisions in service of God and marginalized peoples. By the time my husband, Eric (we met in InterVarsity), and I graduated four years later, we had been pricked by the desire to live counterculturally as urban missionaries in Atlanta. We saw countless biblical examples of God calling individuals and communities to care for the poor, the widow, and the foreigner. It was clear that a big piece of God's heart, and the specific ministry of Jesus Christ, centered on seeking out those society at large often neglects. This spiritual motivation was combined with a deep sense of personal responsibility and desire to live differently in order to help others and glorify God.

After graduation and marriage, we read John Perkins's book *Restoring At-Risk Communities* and Bob Lupton's *Theirs Is the Kingdom.* We prayed and dreamed about participating in Perkins's three Rs of community development—relocation (move to the inner city), redistribution (share resources and invest in the neighborhood), and racial reconciliation. About this time, we connected with another young married InterVarsity couple who were also considering a missional move to the inner city. After one year of shared weekly dinners, we decided to buy our first home together and live a communal life in Southwest Atlanta. The purpose of the communal approach was to have a built-in support network as well as save resources through sharing so more finances could be invested in the neighborhood.

Our first clue that this journey would be difficult came when we tried to get a loan for our new home. Even though all four of us had good credit scores and stable incomes, we had difficulty finding a bank that would make a home loan in our Southwest neighborhood. We were experiencing the lingering effects of redlining (neighborhoods marked "hazardous" in red ink on maps drawn by the federal Home Owners' Loan Corp. from 1935 to 1939) that still disproportionally affect low-income, minority neighborhoods today.[2]

But we finally got the financing, and in May of 2007, the four of us filled two enormous U-Hauls and drove from Buckhead (the nicest side of town) to Southwest Atlanta (one of the roughest zip codes). We didn't know it at the time, but as we drove just a few miles we were crossing through zip codes that differed thirteen years in life expectancy. Buckhead is an affluent area with high-end luxury housing and retail centers where 71 percent have white collar jobs and 75 percent have a bachelor's degree or higher.[3] Our new neighborhood was plagued by high poverty levels (18 percent living at the most extreme level of poverty as measured by the federal government), 55 percent unemployment, 48 percent without a high school degree, and high rates of violent crime.[4]

We arrived full of good intentions and community development training, but had no concept or language yet for the structures affecting lifespan and quality of life for our new neighbors. We stepped out of the moving van and were greeted by, "Huh, we thought we ran white people out years ago." We didn't even realize that two preceding generations ago our neighborhood, and many other Atlanta inner-city neighborhoods, were wealthy, white communities and had experienced white flight to the suburbs.[5] Racial tensions and mistrust had been building for decades, and here we were—the new and only white people in the neighborhood. Despite the fact that we stuck out like a sore thumb, our neighbors generally welcomed us and treated us with kindness.

But at first no one really understood why we were there. Speculation abounded, and people assumed at first that we were undercover cops, which put them on their best behavior when we were around. After months of our insisting this was not the case, they labeled us as hippies and tried selling us a variety of illicit drugs. When we didn't buy the drugs, our home became a primary site for stealing. In the first two years we experienced over eighteen attempted break-ins, most of them occurring shortly after we left for work in the morning. But more importantly, we observed two of the primary neighborhood economies: drug trade and theft. Additionally, with high unemployment and incarceration rates, there was an abundance of desperate parents trying to earn a living. The first time we cut our grass the lawnmower was stolen out of our front yard, and we had to buy it back from the local flea market for $25. My uncle jokingly referred to this as a "creative recycling program" in the community. But in reality there was a lot of economic "creativity" going on because even low-wage jobs were out of reach for most of my neighbors.

When we first moved in, we had vowed not to own a gun or have an alarm system, trusting our neighbors indiscriminately up front and wanting to live a "normal" existence in our new home. Over time we relented to add an alarm system, window bars, and two large guard dogs. This was the second scraping of our ideals against the harder realities of our chosen neighborhood. Just the need for these protective measures changed the way we viewed our home and our community. We didn't know who was breaking in, so we became suspicious of everyone.

I knew intellectually that we were putting ourselves at risk moving into a high crime zip code, but our dedication to the move kept me from internalizing how I would really feel living in this environment. The real and perceived danger in the community quickly took its toll, and I spent most of my time at home feeling anxious. Between negative monthly crime reports at our community meetings and rumors of shootings, home invasions, and car theft, I quickly passed the point of concern and went into a chronic stress mode that kept me scared and ultimately sucked the joy out of life. I remember every time I would hear gunshots at night (which was frequent), I would bury my head under my pillow and cry, shivering with fear and wondering if a tragic incident was simply a matter of time. Every sound in the house was a potential intruder. One night I woke to a weird flapping sound. Terrified, I shook my husband, Eric, and whispered that I thought someone might be in our room. I bundled under the covers while he bravely flipped on the light to discover it was simply a large moth throwing itself at our overhead fan. Even after he told me it was fine, I didn't come out from under the pillows, and I didn't feel better. I was so tired of being afraid.

After experiencing this level of stress for a couple of years, I couldn't imagine the negative accumulation of stress for our neighbors who had been in the community for generations. I was already noticing my own psychological health starting to decline, and couldn't help but wonder how families under chronic stress in poor neighborhoods with few resources could possibly cope. I marveled at the fact that my neighbors were holding their lives together as well as they were, considering their circumstances.

There was another major unraveling. Our small Christian community of four, for which we had high hopes, didn't pan out like any of us thought it would. We were in different stages of life and had incompatible personalities and ways of dealing with stress and conflict. For all the wonderful spiritual things we heard about living in Christian community, it turns out to be ridiculously hard for two young families to co-own a home. We argued about everything, from whether to replace our stolen AC unit to what color garden hose we should buy. Two years in we found our house community imploding in a way that counseling couldn't repair. Just like many of our neighbors, we had too many people living under one roof, and it caused severe relationship dysfunction for everyone involved. It was messy, but we struck a deal and the other couple moved out. Once they left, Eric and I looked around and wondered what to do next.

Another challenge we had not anticipated was the impact of small inconveniences accumulating over time. Because we lived in a food desert, had regular police incidents at local businesses, and lacked neighborhood amenities, we drove outside of the zip code for everything: groceries, date night, gas, clothing, coffee, and so on. I couldn't help but notice that whoever was making decisions about where to locate quality establishments (e.g., grocery stores and restaurants) had obviously decided they didn't belong near my community. My neighbors adjusted by walking a lot or taking MARTA (public transportation) to get the items they needed, but a simple errand could turn into an all-day activity. Over time the steady drip of challenges welled up into a daily perception that everything was hard.

On top of the normal stresses of being working professionals, our life in the neighborhood began affecting our work. I remember arriving very late to work one morning at Boys and Girls Clubs and getting a look from a colleague. I considered explaining about the violent shooting that happened behind my house the night before that kept me from sleeping, or mentioning the fact that I had to drive across town in rush hour traffic that morning to get gas before arriving at my desk. Instead, I just looked down and walked to my office. I was behind before the day had even started.

So two years in we were already scared and weary. Our community dysfunction and the neighborhood challenges took a major toll. While we knew urban ministry would be difficult, we were completely unprepared for how challenging it would feel to two young idealists raised in the suburbs—a completely and utterly different world. We regrouped again and again, trying our best to love our neighbors and advocate for the benefit of the neighborhood the best we knew how. Eric's work in the community largely focused on mentoring young men, giving them a vision for their education, and helping them develop marketable skills. We transformed our home into something like a Boys and Girls Club, complete with a game room, computer lab, two-story wooden clubhouse in the backyard, and a kitchen stocked with healthy snacks. Yet even with all these resources at their disposal, the kids we loved who lived across the street battled poor grades, skipped school, and only sporadically showed up for weekend tutoring programs. I felt dejected and quickly realized that our best efforts were not enough.

My work in the community focused more on fundraising for capital improvement projects at the neighborhood park and lobbying local government officials for resources. There were certainly beautiful "breakthrough moments," like the day our middle school-age neighbor, Caleb, was baptized or the ribbon cutting ceremony for the new pool, playground, and splash pad at our park. But while we saw glimpses of progress, the overall status of the community remained the same, and the family situations largely stagnant. We were doing everything the community development books and training had taught us, but we were underwhelmed with the results. The neighborhood needed so much more than one family or even one organization could provide. The issues were bigger than us, and the cumulative effects of poverty and de-investment in the neighborhood would not be reversed quickly.

This reality sunk in somewhere at the turn of year three. Discouraged, we started brainstorming how we could move out. Unfortunately, we bought our home right before the housing crash of 2007, and it took ten years to regain its value. We spent literally hundreds of hours discussing how to move out without obliterating our finances. It took seven years to finally get to the place where we could leave. In addition to lacking a healthy neighborhood environment, we lacked choice. We were working hard and weren't making bad decisions, but the opportunity to move out was simply not there. Just like many of our neighbors, we weren't in Southwest Atlanta because we loved it—we were stuck there.

I think that was the final nail in the coffin, which most contributed to us becoming mentally sick and critical, mere shadows of our former Christian selves. We spent countless evenings with our heads in our hands at the dinner table asking God, "Where did we go wrong? We wanted to do something important with our lives, why did it turn out like this?" Over time the questions devolved into confusion and despair, "Is God even real? If so, does he just like to see us suffer?" Finally, there was a deep sense of grief and identity crises as we asked, "Who are we, and now what?" In these moments we expected our religion to provide us with answers to what was, in reality, a poorly planned major life decision based on sentiment and inadequate scientific data to support that the community-development model yields results.

When those answers didn't come easily, we turned on each other, replaying past decisions and pointing fingers desperately trying to find someone to blame. We felt abandoned and foolish, proving everyone right who told us not to move there in the first place. We started spending as much time out of the neighborhood as possible in nicer parts of town or vacationing with family for long weekends. But even though we could get away temporarily, we would inevitably come back and be hit in the face by the contrasting harsh realities of life in our neighborhood. One night after spending a lovely evening on the east side of town, we pulled up in our driveway and couldn't even bring ourselves to get out of the car. Instead, we just sat there talking about our current problems, which in turn escalated into yelling—me yelling out of fear for my and our daughter's safety, and Eric yelling in bewilderment on how to provide safe options for his family. The yelling gave way to sobbing, and then we just sat quietly listening to the dripping sound of our tears.

A few weeks after that raw and painful night, I had a vivid nightmare. I dreamed that a group of witches, hair flowing wildly, robes wafting in the wind, were pushing someone (presumably dead) down our street on a stretcher. The sound of the metal stretcher scraping the broken asphalt, coupled with their crazy cackling laughter, was deeply disturbing. It was almost as if the witches represented the harsh environment adding another sick soul, beyond the point of cure, to their bounty. I woke terrified of being the next body on a stretcher. I told Eric we had to get out, no matter the cost. After all was said and done, we realized we wanted and needed the same thing everyone else desires—a safe, healthy place to live and raise our family.

In the process of trying to make sense of it all and rebuild our lives, I tried a thought experiment. I decided to believe without question that God is good and loves me like a kind father. I repeated the truth to myself every day that our lives were valuable to God and that there was a purpose to everything we experienced. It was from this place that I resumed asking him why Eric and I had spent our twenties failing at community development. At the time, I had been working for the Good

Samaritan Health Center for a little over a year and had recently met Keri Norris, chief of health policy and administration for the Fulton-DeKalb Hospital Authority. She introduced me to the startling fact that a person's zip code matters more than your genetic code in determining whether they live or die. Then I found the research from the Virginia Commonwealth University revealing a shocking thirteen-year life-expectancy gap between Atlanta's inner-city neighborhoods and more affluent areas.[6] I realized there was a whole body of scholarship explaining what my family had experienced. What the population health world calls "social determinants" have a huge impact on our health and when sustained over long periods of time lead to toxic stress and ultimately shorten our lifespan and quality of life. In other words, we were not the only ones getting sick in a poor urban neighborhood. The difference for us is that we had health insurance and a strong safety net, while many of our neighbors still suffer from the trauma of poverty and remain untreated.

I am motivated to share my story because if my family, with all our privileges and opportunities, was radically affected by the social determinants of health in less than a decade, I can't help but wonder how our society can expect the poor to "pull themselves up by their bootstraps" and overcome the structures that keep our inner cities in a state of languishing. In hindsight I realize God, in his severe mercy, was allowing my family to experience firsthand how social determinants affect health status. By understanding what is really making poor urban communities sick, we can begin to choose interventions that promote community health and work more strategically to create environments that support health equity for all.

BREANNA'S JOURNEY

I can still picture the church in my central Iowa hometown that housed the newly opened, free medical clinic. As a high school student I had learned of this one-night-per-week clinic from my own pediatric nurse practitioner who volunteered regularly. I convinced the director that I could be helpful, and she graciously allowed me to come every week. So I set off to change the world by organizing the supply closets and registering patients. Over the several years during which I volunteered, I made very little impact in the community, but the experience did everything to change me.

My first memories of being aware of issues of poverty, injustice, and social determinants of health centered around a middle school project on child exploitation. The project grew into an obsession of calling my senator, writing letters to Nike, and boycotting a list of companies so long my parents about lost their minds. I found myself fixated on the injustice of the world and determined to make a career of fighting it. A year or so into this mission I realized I needed some type of skill. How was I going to change the world if I didn't have a tangible skill to offer? The first evening I spent at that small free clinic seemed to answer that question. I watched the providers at the free clinic—family doctors, nurse practitioners, and ER physicians—caring for uninsured patients. I watched them give medication and medical advice, and in my eyes they were superheroes, curing the sick and helping the poor. This was my answer. This was my avenue to change the world.

Interestingly, it was a few years later during nursing school at another free clinic where my visions of changing the world were abruptly cut short. By this time I had some, albeit limited, skills and was working with a team of community members to open the first free clinic in a small college town in Minnesota. It took only a few weeks to realize that medicine was not all that these uninsured patients needed. A course of antibiotics didn't cure poverty. Medications for diabetes didn't seem all that helpful for people struggling to afford food. A consult at the free clinic couldn't make up for missed screenings or needed surgeries only health insurance could buy. I regrouped and educated myself on health policy. By the time I got to Emory University for my master's program, I was splitting my time equally between advocacy at the local and state levels and my schoolwork. Yet the further I researched health policy and the impact of health insurance, the more I realized access to health insurance and health care was only one small determinant of health status.

One month after our college graduation, my husband, Matt, and I moved across the country to start our life in Atlanta. We had a small sized U-Haul filled with a combination of dorm furniture and donations from our parents. We had my old but well-kept Altima and the bright red Camry my father-in-law had graciously leased for us so we'd have at least "one reliable vehicle." We had our first month's rent for a 700 square-foot apartment near campus and a few hundred dollars in wedding money. Matt started teaching first grade in the Atlanta Public School system at a school where nearly 100 percent of students were receiving free and reduced lunches. I started at Emory University to study health policy and become a nurse practitioner. When one of my nursing friends told me about a small clinic serving people without insurance just around the corner from my husband's school, I called the director and begged her to let me work with her for my first clinical rotation.

The clinic was a small, rundown building next to the gas station and had bars on the door and the only two windows in the entire building. Inside were a front desk, bathroom, single sink, and two exam rooms. I introduced myself to the director, who was also the sole health care provider at the clinic. The director had the experience that comes from a long career and the hardness born of a life filled with difficulties. She had moved across the county; she had been homeless; she told stories of being so sick and without a dollar or any insurance and begging a local doctor for a round of antibiotics. Her personal experience shaped her passion to provide health care to those lacking insurance and financial resources. I walked in ready to begin the career I had imagined since high school.

The director peered over her glasses at me and asked, "Are you the one that wants to come work at a place like this when you graduate?"

"Oh yes, this is what I want to do!" I responded eagerly.

She turned back to the towering stack of charts on her desk. "Huh. You driving up in that fancy car; we'll see how you do," she smirked with a voice dripping with skepticism and sarcasm.

At first I felt anger rising inside me, *How can she judge me like that? Doesn't she know how long I've waited for this? Does she know how many hours I have spent in clinics like this?*

But anger was quickly replaced by a sinking feeling that stuck with me for months. *She's right. I have no relevant life experience. I have never been hungry, or poor, or homeless. I've had health insurance and good health care for every illness in my life. I am a fool to think I can do this.*

But I stayed that day and came back the next. Each day I returned, I learned what school never taught me. Prescribing medication was helpful, but it wasn't fixing what was already broken. Diseases weren't making my patients sick. Poverty, stress, trauma, and food insecurity had taken a devastating toll on their health status long before they showed up at my office. The clinical placement became a job at the clinic after graduation and Matt and I bought a foreclosure just down the road. A few years later I started working at the Good Samaritan Health Center, continuing to bridge the health care access gap between those with and those without health insurance. At Good Sam I am a part of an interdisciplinary team providing comprehensive primary care to uninsured families. My work remains a small piece in restoring a vast and devastating inequity in which social status determines health status.

Yet I have also learned resilience from my patients who had lived through experiences much different from my own. I learned that partnerships between patients and providers can be more curative than prescriptions. Most importantly, I have come to understand that good health care alone is not the end but rather a means of highlighting and addressing social determinants of health. Healthier communities are possible, but only when we understand what is making them sick and only when we are willing to reconceptualize health and our role in it.

CHAPTER 2

WHAT IS MAKING US SICK?

AN INTRODUCTION TO SOCIAL DETERMINANTS OF HEALTH

BREANNA

Around the turn of the nineteenth century, a strange new theory was threatening the basic principles of modern medicine. Educated physicians understood disease to be a result of imbalances in body fluids, punishment for sin, or miasmas, bad odors delivering disease into the body. The idea that diseases were contagious was considered uneducated, a belief held by unsophisticated, traditional healers.[1] Despite the prevalence of this belief, a few physicians proposed that perhaps the world was wrong about disease. Hungarian physician Ignaz Semmelweis suggested that physicians who autopsied the bodies of dead mothers might be spreading puerperal fever to women when they then delivered babies without washing their hands. The obstetrician implemented hand washing with chlorinated lime salts. While the incidence of puerperal fever plummeted, he was ridiculed and dismissed from the hospital.[2]

Years later, researchers began to prove his theory. In Germany, Dr. Robert Koch was experimenting with anthrax and mice. By injecting infected blood into healthy mice, Dr. Koch proved that specific germs could cause specific diseases.[3] Meanwhile, in France, Louis Pasteur proposed that microbes in the air contaminated food, causing illness. His germ theory was scandalous, with one newspaper exclaiming, "I am afraid that the experiments you quote, M. Pasteur, will turn against you. The world into which you wish to take us is really too fantastic."[4]

It is difficult to imagine a world before germ theory. As a healthcare provider, it seems impossible to consider a time when washing hands after attending a sick patient was not common sense. And yet the pioneers of germ theory were scorned, rejected, and, in the case of Dr. Semmelweis, even committed to an insane asylum. Germ theory revolutionized medical care. Yet, in the economically developed world, most people aren't dying from germs anymore. The top ten causes of death in the United States, accounting for 74 percent of all deaths, are nearly all noninfectious diseases, with heart disease (first) and cancer (second) accounting for nearly half of all deaths.[5] Many of these chronic diseases are preventable.

Today, the world faces the same questions that led Semmelweis, Koch, and Pasteur to their controversial experiments and revolutionary discoveries: What is making us sick?

Michael Marmot found himself asking this question decades ago as a young physician in England. As he treated patients in the inner-city hospital, he began to question the value of his treatments.[6] It seemed to him their health problems stemmed from factors he couldn't address within the clinic. So he left the clinic in pursuit of an answer and studied epidemiology. He began working with the Whitehall study, a survey of eighteen thousand male civil servants in London who were initially surveyed between 1967 and 1969. He followed the cohort's mortality rates and published his groundbreaking conclusion in 1978. Men with the lowest grade of employment had three to six times the coronary heart disease mortality as men with the highest grade of employment.[7] However, this difference was not only observed between men at the very top and men at the very bottom of the employment gradient. Marmot discovered an inverse association between grade of employment and heart disease mortality. In other words, the higher the man's employment status, the less likely he was to die of heart disease.

As any physician would, Marmot considered known risk factors for heart disease. A difference in risks like diabetes and high blood pressure could easily explain the finding and support what was already known about the causes of heart disease. Indeed, men at lower levels of employment did have higher blood pressures, heavier weights, and smoked more. However, when Marmot controlled for all of these medical risk factors, his findings remained unchanged. The increased risk of heart disease mortality among men of lower status still existed. Even more troubling is that this cohort of men shared other significant similarities. They were all employed (as civil servants), worked in the same city (London), and by virtue of employment and location, enjoyed similar access to health care. This study wasn't comparing rich CEOs against the unemployed and homeless. All of these men were civil servants. If cardiac risk factors could not explain the findings and neither could the presence of employment or access to health care, what did? Marmot had discovered the social gradient of health. Social determinants like income, social status, type of employment, housing, and education correlate to changes in health status along this gradient.

What Marmot discovered in the 1970s explains why people in West and Southwest Atlanta neighborhoods die thirteen years earlier than those in wealthy neighborhoods. It also explains why by the time people arrive at the clinic to see me, my interventions do little to decrease this risk. The social gradient of health simply states that where a person is on the social ladder determines their health status and mortality. I imagine at the turn of the nineteenth century the idea of germs spreading from one person to another and causing disease was more terrifying than helpful. The same can be said for the theory of social determinants of health. The World Health Organization defines social determinants of health as "the conditions in which people are born, grow, work, live, and age" and the "forces and systems," including economic, social, and political, that shape the conditions of daily life.[8] This means that the conditions of our neighborhood, our jobs, our education, and the social structures surrounding us have a direct impact on health status and life expectancy. People who live in disadvantaged neighborhoods are more likely to have a heart attack than people who live in middle-class neighborhoods, even when income is taken into account.[9] If social determinants such as poverty, education, living environments, and social status are contributing to disease, what are we going to do about it?

While it was one thing for Semmelweis to propose germs could be transmitted from diseased dead bodies to healthy women, germ theory was proven when Pasteur infected mice with rod-shaped anthrax bacterium. Today, it is widely understood that people get viruses and infections through the spread of germs. The person at the grocery store coughs into her hands while pushing the cart. She leaves her cart in the corral just as I pull in and grab the cart handle covered with her germs. I rub my eyes, itch my nose, or forget to wash my hands before eating, and her germs enter my body. If my immune system cannot address the exposure, I am going to get sick. Similarly, it may seem intuitive that poverty, homelessness, and unemployment could make a person more likely to get sick; however, do these social determinants actually cause disease? Decades after the Whitehall study proved the existence of the social gradient, recent research is now explaining the pathology behind it. It turns out social determinants of health act like germs on a more chronic level. Chronic stress from lower societal status, poverty, unemployment, violence, and other factors enter the body and begin making permanent changes.

The term *allostasis* describes our ability to stay the same through stress.[10] Our bodies are built to handle stress and have protective mechanisms that allow us to remain intact during moments of extreme emotional or physical stress. However, we are not all equally equipped with these positive mechanisms. Genetic factors and developmental events such as childhood poverty or abuse can cause an individual to develop ineffective mechanisms when responding to stress.[11] For some, allostasis is much more difficult to obtain. In addition, the body wasn't designed to handle constant bombardments to normal functioning. Allostatic load refers to the wear and tear on our bodies when we undergo repeated cycles of allostasis.[12] Chronic stress from poverty, unemployment or job strain, community violence, and food insecurity increase this allostatic load. This chronic stress activates autonomic, neuroendocrine, and immunological responses.[13] Over time, social determinants are literally altering the nervous system, endocrine system, and immune system within the individual. Normal system functioning turns to maladaptive processes wreaking havoc throughout the body. Basically, the body is trying to keep up but is inefficient and eventually unable. The result is medical problems such as premature birth, cardiovascular disease, and hypertension.[14]

When I sit across from my patient with high blood pressure or a newly pregnant teen, I am working against a gradient established since birth. I am up against a body system that has been overrun by factors well outside the scope of medical practice. It's not that social determinants make some people more likely to get sick than others, as I had initially assumed. Social determinants actually have a direct influence on disease, with consistent, strong associations between social factors and health outcomes.[15] Imagine an epidemic hitting your neighboring county. People are dying and the treatments aren't working. Living five miles from the neighboring county wouldn't be very reassuring. Yet five miles makes a life or death difference in many US urban areas, as shown in the Virginia Commonwealth University Life Expectancy Maps. This is no longer a question of health care access or appropriate medications. Clearly, something much more ominous is behind this tragedy. We are missing something.

Ignoring social determinants is equivalent to treating patients with unwashed hands and wondering why disease keeps spreading. Providers are already considering the way our understanding of social determinants affects clinical practice. If we are going to successfully treat diabetes, we need to assess for food insecurity. If we want to prevent the health effects of chronic stress, we need to assess patients' socioeconomic status and housing situation. Improving health status requires opportunities for education and career advancement, stable and safe living environments, access to healthy food, and environments that promote social cohesion. This will require policy changes, innovative programs, and new approaches to health care. It also means that medical providers alone cannot make people healthy. Community members, city planners, social services providers, educators, faith communities, and businesses must collaborate to create a new normal: a society where zip code doesn't determine life expectancy.

Given the size and gravity of the topic, it can be helpful to consider a framework for understanding social determinants. One such framework is the place-based organizing framework established by the Department of Health and Human Services (DHHS). The definition of social determinants of health used by DHHS mirrors that of the World Health Organization. These determinants are conditions (social, economic, and physical) "in which people are born, live, work, and age that affect their health."[16] The five key areas within the organizing framework include economic stability, education, social and community context, health and health care, and neighborhood and built environment. These areas both overlap and influence one another. Our chapters are structured to address specific components of each of these five key areas.

Economic stability includes poverty, employment, food insecurity, and housing instability. In chapter three we discuss causes of poverty and the ways that poverty affects health. Chapter four discusses employment, considering the health impact of work as well as the effects of unemployment. Chapter five addresses food insecurity as well as health behavior related to nutrition and exercise. Chapter seven reviews the affordable housing crisis and the health conditions resulting from unsuitable living environments.

Education considers early childhood learning, literacy, high school graduation, and higher education. Chapter six considers the impact of early childhood education and environment as well as correlations between high school graduation and health.

Social and community context includes topics such as social cohesion and discrimination. In chapter seven we discuss difficulties of cohesion in stressed neighborhoods, and we address racial and age-related disparities in social determinants of health in chapter three.

Chapter eight focuses exclusively on *health care*, discussing health insurance, access to health care, and use of primary care. We discuss the health disparities resulting from lack of equitable health care access.

Neighborhood and built environment includes food access, quality of housing, crime, and environmental conditions. We discuss food access in chapter five and housing quality in chapter seven. We also focus particularly on the home environment in chapter seven and work environment in chapter four.

Each of these areas could be a book in and of themselves. Here we present merely a starting point for understanding our nation's declining health and how local interventions can bring healing. In part one we will share our personal stories to describe how social determinants are impacting our health and the health of our community and patients along with emerging research on social determinants of health. Our stories involve friends, coworkers, neighbors, and patients who have shared their experiences with us. We strive to tell their stories as accurately as possible and have changed names and specific details to protect their identity. A vast amount of research exists on the subject of social determinants of health, and we have cited many experts, research, and pioneering work in the chapters ahead. We share our stories of struggle, failure, and learning, while recognizing that our stories are simply that and do not represent the lived experiences of everyone in our communities.

In part two we discuss ideas and models for creating healthier communities through social-determinant-focused approaches. This includes our own experience at the Good Samaritan Health Center as well as creative programs from across the country.[17] Despite our frustrations and setbacks in our work and community, our experiences have also convinced us that a healthier world is very much possible. Health equity is the opposite of a social gradient in which health is largely influenced by social determinants. Health equity means that every person has a fair and just chance to be healthy.[18] Obtaining health equity requires the removal of oppressive structures like racism and classism and creating opportunities for better jobs, quality education, safe environments, quality health care, and affordable housing. Equity is differentiated from equality in that equality describes everyone receiving the exact same amount of resources. However, providing the same resources to everyone cannot reverse the social gradient or bridge the thirteen-year life-expectancy gap. Equitable provision of support is not equal but rather fair and just. As depicted in figure 1, an equitable approach to health requires us to provide specific types of support to people based on their need, giving them all a fair chance to be healthy.

In figure 1, the tall individual represents someone with societal advantages such as sufficient income, access to quality education, and healthy family dynamics. This individual will still need support at times to live a healthy life. For example, he may need extensive health care if he contracts a serious illness, or maybe extra tutoring in math during high school. The shortest individual represents someone of lower social status who may have been born into poverty, experienced

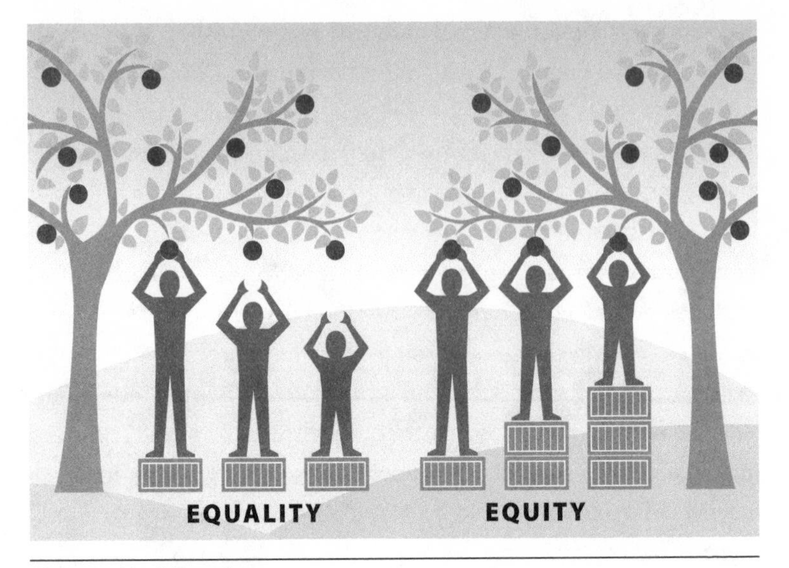

Figure 1. Health equity

childhood trauma, or experienced oppression. This individual will need additional supports to address these social determinants that have the potential to negatively impact health. He needs more boxes to reach the same health status as the taller individual.

Achieving health equity can add years of life to people living on the lower portion of the social gradient. However, regardless of where you are on the social gradient, equity matters. As people on the lower rungs get sicker, all of society suffers. Advancement, growth, innovation, and wealth require a nation of people who are healthy. As the toxic effects of stress from poverty, unemployment, substandard housing, and food insufficiency rob more people of the opportunity to live a healthy life, the health of our nation declines.

Equity also has importance for faith congregations. Equity is a key attribute of God's character, and the word appears in the Bible over twenty-three times. It is the English translation of several Hebrew words meaning "justice," coming from a root which means "straight" (uprightness) or "level" (as in a level playing field).[19] Equity was established by God from the beginning as the goal of a well-functioning, fair, and righteous society. Consider just one example from Scripture: "Thy throne, O God, is for ever and ever: / A scepter of equity is the scepter of thy kingdom" (Psalm 45:6 ASV). It isn't so much that equity is good just for equity's sake. Rather it is an important part of God's character. God is righteous and just, judging with fairness. Isaiah 11:4 says, "But with *righteousness* he will judge the needy, / with justice he will give decisions for the poor of the earth." Advocating for health equity in society is in alignment with God's heart and when practiced serves as a witness to the world of God's kindness and fairness toward humankind. God's vision is a kingdom without pain, sickness, or mourning (Revelation 21:1-4). Healthy, equitable communities embody the kingdom of God in this world.

Creating healthy communities requires initiatives and policy change at local and national levels to achieve health equity. We highlight some of this work throughout part two in addition to a brief discussion on effective activism. We advocate for policies that promote health equity, regardless of political party. National, bipartisan commitment to narrowing life-expectancy gaps and creating healthier communities is necessary to achieve health equity. We also feature local level initiatives that are actively improving the lives of community members. Our hope and prayer is that we, along with health care providers, urban pioneers, social-service practitioners, community advocates, and faith communities across the country, can challenge existing structures that are making our neighborhoods sick and ultimately restore the health of our communities.

CHAPTER 3

THE TRAUMA OF POVERTY

VERONICA

After living in the midst of it, the best way I can describe poverty is like a leaking faucet. Certainly, there can be major traumatic moments for families living in poverty, but for most it is the slow, steady drip of thousands of difficult moments that most affects health. According to Eric Jensen, in his article "The Effects of Poverty on the Brain," sustained poverty has the same impact on the human brain as a traumatic event. In other words, poverty is itself a trauma.[1]

I can see how this is the case. Even though Eric (my husband) and I are a dual-income middle-class family with a strong support network, our experience of living in an impoverished zip code took its toll on our mental health. We never experienced any large, personal crisis, just the unrelenting drip of difficult moments, with no time to recover. It's not that we weren't resilient, but we never got a break. We spent all of our time unsuccessfully trying to stem the tide. For example:

- Our housemates moved out and now our mortgage payment is doubled, leaving us financially strapped (drip).
- Our AC unit stopped working because someone cut the Freon line; it's the middle of summer and it's expensive to fix (drip); our utility bills are high because our house is one hundred years old and poorly insulated (drip).

- The alarm company just called, our crawl space was broken into again (drip). We meet the police at the house to assess the situation. All of our quality tools have been stolen, and the thief went to the bathroom in our basement (drip).
- Gunshots are fired, scaring the dogs, and they dig out and are loose in the neighborhood (drip). I leave work in the middle of a meeting to come home and retrieve them (drip). On my way back to the office I accidently cut someone off and get chewed out with a string of racial slurs (drip).
- While trying to enjoy the front porch swing, we are eaten up by mosquitos because our neighbor's yard is overgrown and has pools of standing water (drip).
- A dump truck is on the dirt road behind our house, leaving a mountain of tires and trash behind (drip).
- Rodents are rustling in the attic again, despite repeated service calls (drip).
- We're tired at the end of a long workday and want to order food or run to the grocery store, but no healthy options exist and we don't have the will to drive across town in rush-hour traffic. We eat cereal for dinner, again (drip).
- A new stretch of paved walking trails opens a half-mile from our house, but the walking path to get there is full of harmful advertising, broken sidewalks, and busy streets on which pushing a baby stroller is unsafe. We eventually give up on walking in our neighborhood altogether (drip).

Intense, prolonged stress often manifests itself differently in men and women. Experts have found that residents living in neighborhoods with poor housing stock experience high blood pressure, increased levels of stress, and more symptoms of depression.[2] In our case, I developed major depression and Eric had regular panic attacks.

This led to a debilitating cycle where something bad would happen in the neighborhood and I would descend into the depths—finding it difficult to get out of bed or manage my life. Eric, in his concern for me and feeling helpless to change the situation, would get angry and resentful at God—sending me back into sadness and hopelessness. We cycled through this discouraging process for a long time. It wasn't until one of the neighborhood kids noticed something was wrong that I realized I needed help.

I had been spending more time in bed crying and was right in the middle of a crying spell when one of the neighborhood kids, Caleb, knocked his loud, uninterrupted knock. I didn't check my reflection or think about my appearance until I opened the door. Looking shocked, Caleb said, "Uh . . . Veronica . . . What is wrong with you? Have you been crying?" Embarrassed I lied and told him I was sick and must have smudged my makeup, then sent him home. After closing the door, it dawned on me that I needed to see a counselor, to which I had access thanks to my health insurance and income level. But for so many in my community who were undoubtedly reeling from the effects of chronic stress on the brain over a lifetime, or over generations, mental-health treatment was not an available option.

When so many things are wrong, really wrong (not just wrong in your head) seeking mental-health treatment feels like a Band-Aid that won't change any of the heavy issues stacked against you. In our case, neither of us had struggled with mental-health issues before Southwest Atlanta, rather our bodies and minds were reacting to the hard situation. We were experiencing what the scientific community refers to as epigenetic effects. Epigenetic effects do not touch or change DNA, but rather install something like a switch that can turn existing genes on or off and influence health-related factors like inflammation, the immune system, chronic disease, and so on. As researcher J. Stern puts it, "Genes load the gun, but the environment pulls the trigger."

In a February 2016 article titled "Mental Illness Is a Result of Misery, yet We Still Stigmatize It," author Richard Bentall points out that while genes play some role in psychiatric conditions, recent research tells a more complex story. He concludes that the biggest cause of human misery is miserable relationships with other people conducted in miserable circumstances.[3] Consider this description of poverty from a 2015 article by the Kaiser Family Foundation:

> In the United States, the likelihood of premature death increases as income goes down. Similarly, lower education levels are directly correlated with lower income, higher likelihood of smoking, and shorter life expectancy. Children born to parents who have not completed high school are more likely to live in an environment that poses barriers to health. Their neighborhoods are more likely to be unsafe, have exposed garbage or litter, and have poor or dilapidated housing and vandalism. They also are less likely to have sidewalks, parks or playgrounds, recreation centers, or a library. In addition, poor members of racial and ethnic minority communities are more likely to live in neighborhoods with concentrated poverty than their poor White counterparts. There is also growing evidence demonstrating that stress negatively impacts health for children and adults across the lifespan. Recent research showing that where a child grows up impacts his or her future economic opportunities as an adult also suggests that the environment in which an individual lives may have multi-generational impacts.[4]

While everyone's life includes a certain degree of misery, it's turned all the way up for those living in poverty. On one particularly miserable day, the vacant house next door to ours caught fire. This home was in terrible condition and had been a flagrant code enforcement and health issue for years. The neighborhood had been experiencing a string of arson incidents, and on this day someone set our neighbor's house on fire.

I was in the third trimester of my first pregnancy the day of the fire and the firefighters, in a desperate attempt to put the fire out, were furiously throwing random items out of the house, breaking our guest bathroom window in the process. I had spent months "nesting" and preparing everything for my daughter's arrival—only to have my side yard, fence, and bathroom window destroyed in less than five minutes. While I was grateful to the firefighters for protecting our house from further damage, my eyes welled up with tears as I looked at the mess that was now my front yard. As the firefighters were packing up, they asked us why we lived there and pointed out that we had bought a house in a really rough part of town. When I finally got back inside and sat down, super pregnant and super uncomfortable, the tears dripped heavy on my growing belly.

After the fire incident, Eric started making calls to the city because the house was now a more severe code enforcement issue, and we reasoned that no one would ever buy our home as long as this persisted. We have documentation of hundreds of calls and emails Eric made over several years about this specific case before it was resolved. In addition to lacking the amenities of the suburbs, we felt ignored even on the most basic city services, frustrated and powerless to change our situation. This is a common reality in poor neighborhoods. Case studies demonstrate how community voice and civic engagement are critical for successful place-based strategies. However, that voice is often muted due to lack of local leadership that can connect to policy change efforts. In other words, there is no one living in the community with the influence to pull levers for sustainable change at the local, state, and even national levels. Furthermore, weak social and political networks make it difficult for communities to organize against toxins that are located in their neighborhoods.[5]

When I think about people dealing with similar issues in the inner city but also the added stressors of extreme financial strain, lack of food and clothing, unsanitary living conditions, or intense family dysfunction (such as domestic abuse or drug activity), the cards seem completely stacked against them. In our case without health care access and capable counselors, I don't know if we would still be here today. On more than one desperate occasion Eric and I came to the conclusion that life was simply not worth living. Whereas in a normal life there are ups and downs and "you win some, you lose some," in our Southwest Atlanta neighborhood it felt like the news was always bad. The kids are skipping class and doing drugs, the dads are absent, the ATV gang woke the baby up again, the neighbor is getting evicted, the bill is past due (drip, drip, drip). We found we could only take in so much "bad news" before starting to feel like life was absent of any goodness and it would be better to die. I have come to learn that what we were experiencing is commonly referred to as "toxic stress" and is the "prolonged activation of the stress response system that can disrupt brain architecture and other organ systems, and increase the risk for stress-related disease and cognitive impairment."[6] Trouble and despair were literally changing the way we viewed life.

If we want to help our inner-city neighbors have better outcomes when it comes to holding down jobs, graduating, parenting, and so on, we must address the significant mental-health needs that result from poverty. In fact, many leading philanthropic foundations have recently restructured their giving philosophies to start with mental-health funding. They now realize that the societal improvements they seek start with the mental well-being of adults and children, and have shifted their contributions accordingly.

In my community development training, I learned the importance of "moving in." As a means of following the biblical example of "the Word became flesh and made his dwelling among us" (John 1:14), we were to move in to the neighborhood and bring God's love and advocacy for positive change. But if I am honest, if you had come to me when I was at my worst in Southwest Atlanta and said, "Good news! Some new people are moving in to the neighborhood to love the community!" I would not have been happy. I wanted real, practical, immediate relief from the social determinants that were bearing down on my family. I didn't need more friends—I needed a psychiatrist.

I took a single psychology class in college, and one concept I never forgot was the rainbow-colored pyramid of Maslow's hierarchy of needs. Having grown up in a nice, suburban neighborhood, I only knew people whose needs were met most of the time. The idea that you needed full access to basic needs in order to progress to higher-level needs like self-esteem was new. Living in Southwest Atlanta, especially in the early years, I recognized the reality of Maslow's hierarchy of needs. Health problems abounded, some chronic and acutely painful, and went years untreated. Safety was elusive and varied greatly depending on what block or street you lived on. Love and belonging were strained and family relationships thorny. In some of his writings even Maslow admits, "The higher-order (self-esteem and self-actualization) and lower-order (physiological, safety, and love) need classification is not universal and may vary across cultures due to individual differences and availability of resources in the region or geopolitical entity/country."[7]

So there is a problem. Progression up Maslow's hierarchy is not available to everyone. Wess Stafford writes in *Too Small to Ignore* that

> at its very core, poverty is a mind-set that goes far beyond the tragic circumstances. It is the cruel, destructive message that gets whispered in the ear of millions by the enemy Satan himself: "Give up! You don't matter. Nobody cares about you." . . . More than any other emotion, the poor feel overwhelmed. Without financial resources, shelter, food, education, justice, or skills to address their plight, they succumb to the downward spiral that leads to hopelessness and despair. That, my friend, is the essence of poverty.[8]

BREANNA

Poverty is a strong predictor of mortality and a critical social determinant of health. The Department of Health and Human Services lists poverty as a key component of economic stability. Being poor literally shortens your life span. When considering race, gender, employment, neighborhood socioeconomic status, and household income, household income is the strongest predictor of mortality.[9] Those in the lowest socioeconomic bracket have a threefold higher mortality than those in the highest socioeconomic bracket. Poverty is also not equally distributed throughout society in terms of risk and health impact. Disparities in incidence and impact are particularly severe for children and racial minorities.

One in five children live in poverty as defined by the federal poverty level, and 41 percent live in low-income families.[10] The stress associated with lower socioeconomic status creates biological and physical changes in the body, resulting in higher risks of disease and mortality. While detrimental in adulthood, these changes are devastating in childhood. Health in early life has lifelong effects on health status.[11] The biological changes due to poverty in childhood limit developmental capacity.[12] Poorer socioeconomic conditions during childhood cause an increased risk of cardiovascular disease in adulthood irrespective of adult socioeconomic status.[13] In other words, your life expectancy not only depends on your current zip code but also the zip code where you were born.

Childhood socioeconomic status becomes even more important for children with health problems, learning disabilities, or other diagnoses. Dr. Jen, a pediatrician who runs Good Sam's Developmental Clinic, remembers the experience that prompted her to initiate the developmental clinic, along with a multidisciplinary team, to provide family assistance and health care for children with developmental disorders.[14] Her son was diagnosed with autism prompting a personal journey to identify any resources that might assist him. "I had the ability to quit work," she explains. "I could take him to speech therapy and occupational therapy." She started volunteering at Good Sam and one day entered the exam room to find a mother sitting with her young son working on flashcards. The boy had been diagnosed with autism, but his mother was on her own to assist him. As a single mother working full-time she did not have the time or financial resources to supply him with additional support. "I saw this mom working with her kid the way I would be with mine," Dr. Jen remembers. "Yet her resources were so much more limited than mine. It seemed so unfair that my child had a chance for a better outcome than her son." The difference was not in their love for their children, work ethic, or dedication to providing the best they could, but rather a difference in socioeconomic status.

Poverty is also more prevalent among racial and ethnic minorities. Twenty-four percent of black Americans and 21 percent of Hispanic Americans live at or below the poverty line compared to 9 percent of white Americans.[15] Health disparities mirror these statistics. Black Americans have a higher death rate than white Americans for eight of the ten leading causes of death, and black infants are more than two times more likely to die than white infants.[16] Segregation lies at the heart of health disparities as it shapes the socioeconomic conditions, education, and employment options at individual and community levels.[17] Racism has a direct negative impact on health. A review of over one hundred studies found that experiencing racism was consistently associated with negative mental-health outcomes and health-related behaviors.[18] Improving health outcomes depends on dismantling systems of oppression that allow racism and segregation to persist in the United States.

The difficult reality is that systems that so devastatingly impact the lives of some, benefit the lives of others. Most people don't want to see their neighbors suffer, and reading about the gross injustices within our wealthy nation is uncomfortable at best. Social determinants of health have largely been favorable to me as an upper-middle-class, well-educated white woman. The social constructs bringing privilege to me are working against the patients I see every day. Growing up, all of my health care providers looked like me. I never questioned whether I could join that profession someday. As a teenager, people asked me where I was going to college rather than what I was doing after graduation. My public school had ample resources, and my home had books filled with characters I could relate to. Racism and classism have worked to my advantage. This is an unsettling reality. As most people in poverty are not there as a direct result of their own actions, those on the upper rungs of social status did not choose comfort and longevity at the expense of others. Regardless, it does not absolve us from taking responsibility for this reality.

As a society, we prefer situations that are win-win. We desire for the lives of others to be better, but not at the expense of our status and comfort. The problem lies not in whether or not such solutions exist, but in our definition of *win*. As I look at my life and work in the face of devastating injustice, I have a choice to start with the question, What will make my neighbors healthy? I have to be willing to accept that the answer might not be comfortable.

In Bryan Stevenson's *Just Mercy: A Story of Justice and Redemption*, he states that the opposite of poverty isn't wealth, it is justice.[19] The gross disparities in health status and life expectancy in this country are symptoms of a deep, unrelenting injustice.

CHAPTER 4

WORKING TO DEATH

EMPLOYMENT AND SOCIAL STATUS

Not enough to live on, but
a little too much to die.

MIKE MILLIUS, SONGWRITER

VERONICA

We are all huddled around the computer, scanning job descriptions and chatting with Brandon as he applies for work. This had been our daily routine for months, with a few fits and starts, but so far no job. Brandon was trying his hardest to leave a life of crime and secure honest work, but he had a criminal record and no real marketable skills. Eric and I used our best creative thinking to help him draft a résumé, and we practiced interviewing and lent him some professional clothes. Eric had written letters to judges over the years to request community service instead of a sentence for Brandon, all in an effort to move him in the direction of restoration, not incarceration.

All of a sudden, Brandon smiles and says, "Do you guys want to see pictures of my family members?" We laugh and nod, grateful for a break from the slog of applying for jobs. Before we can process what's happening, Brandon goes online and starts Googling names. We see the first mug shot—it's his dad. Then the second mug shot, his uncle. This continues for another two or three family members, all of them still in jail serving time. Brandon, a big and normally jovial guy, pauses for a moment and says very seriously, "I love these guys, but I don't want to end up like them."

There's a song on Lecrae's *Rehab* album that always reminds me of this moment. It's called "Just Like You" and tells the story of a young boy looking up to uncles who loved him but were involved in dangerous lifestyles. This song reminds me of the generational nature of poverty and crime, that young boys seeking attention will copy what they see older men in their lives doing—even to their own harm. Yet the song also talks about the fondness and familiarity that is established in these generational patterns, and gives a glimpse into the blurry space between victim and perpetrator:

> You showed me stuff I probably shouldn't have seen,
> But you had barely made it out your teens,
> Took me under your wings
> I wanted hats, I wanted clothes just like you,
> Lean to the side when I rolled just like you.[1]

In Brandon's case, even though he was being brave and bucking the neighborhood trend by applying for honest work, it was hard. Nonetheless, Brandon was determined. He had a daughter and a son on the way and wanted to provide a different future for them.

So Brandon would wait until we got home and sneak over to search Monster.com or the local Taco Bell employment page. He didn't want "his guys" to know about this because even though he was a leader, this wasn't cool. So we filled out tons of job applications, bought countless MARTA cards so he could attend interviews, gave him rides, bought him nice clothes, wrote and rewrote cover letters and résumés, and practiced interviewing (and smiling while giving a firm handshake). In the end, his record kept him from getting hired, and he went back to the life he knew—selling drugs and stealing for a living. Phil Hunter, executive director of Georgia Works, an organization helping homeless men find permanent work, describes the difficulty of finding employment with any type of criminal history: "On their own they wouldn't get through the front door."[2] Yet, men like Brandon aren't unemployed because they are poor workers.

"People leaving a life of drug-related crime are very resourceful folks," says Phil Hunter. "Get them a job and they are great employees." Georgia Works advocates for men like Brandon, giving them the chance employers might not otherwise. In Brandon's case, employers couldn't see past his record. Brandon needed to buy food, clothing, and save enough so he and his little family could move from the overcrowded duplex they were living in. Brandon knew that moving to a better neighborhood would be healthier for his family, and in pursuit of this goal he returned to his former strategies of earning money.

Not long after going back to his former lifestyle, Brandon ended up serving time at a county jail for participating in an armed robbery. One night we drove Aliyah, Brandon's teenage girlfriend, to the jail. We were visiting because Aliyah had given birth to his son while he was in jail and wanted to introduce him to Brandon Jr. I watched with tears in my eyes as Aliyah spoke to Brandon through a phone, watching him on a television screen (in a bright orange jumpsuit) while trying to hold up the newborn so he could meet his son. Brandon was so proud and kept saying how much he wished he could get out and be the father his son deserved. Certainly, Brandon had made some major mistakes, but it was complicated. My previous framework for classifying "victim" and "perpetrator" was totally blown away, and all that was left was a sweet little family that we dearly loved caught in a vicious cycle of generational poverty and trying to do their best with what they had.

Jail is not the only, or even the most, severe consequence Brandon and his family will endure. Black men with less than a high school education have a life expectancy of over fourteen years less than highly educated white men born the same year.[3] Brandon's chronic unemployment is changing his health and the health of the family he so desperately wants to protect. Most studies find that unemployment is associated with poorer overall health; however this is particularly true with regard to mental health. When compared to employed individuals, unemployed people report much poorer mental health and lower well-being.[4] Those who are unemployed are also less likely to be insured and more likely to delay health care due to costs.[5]

Unfortunately, employment isn't always the solution. Brandon's story illustrates another important reality: a low-wage job cannot support a family. Brandon couldn't earn enough to support his family working a low-skilled, minimum-wage job. He is not alone. There is not a state in the United States where a person working full-time at minimum wage can afford a two-bedroom apartment.[6] What happens when employment isn't enough?

Recently I had lunch at a nice bistro on the north side of town. As I savored a pita sandwich I couldn't help but notice the young black man who was bussing tables. He had a winning smile and worked diligently, graciously serving the customer base, which was all wealthy, white families. I marveled not only as his work ethic but his kindness toward those of a different race and class. I thought about that young man for the rest of the afternoon, remembering Brandon's experience and wishing he could have been given a chance to shine.

A few days later I was walking into a Starbucks on the west side of town and saw him. The same young man with a great smile was sitting on a public bench, now dejected. I did a double take as I walked by, trying to determine if it really was the same guy. Surely not, because why would he be on this side of town during work hours? After buying my latte I walked by again, and this time curiosity got the better of me.

"Excuse me sir, but do you work at the Zoe's on the north side?" I asked.

"Yes," he smiled and shook my hand.

"Well, forgive me for staring, but I thought it might be you. I ate lunch there last week and couldn't help but notice what a good job you did. Thank you."

He smiled again, but this time there was sadness behind it. He went on to explain that after work that day he was smoking in a public park and got arrested, spent three days in jail, and had just been released and had no money to even catch the bus to work to see if he still had a job. I was shocked. Before I could ask clarifying questions, he pulled out his ticket, and sure enough the only crime noted was that he smoked in a park. It seemed inconceivable to me that this black man could be arrested in handcuffs, taken to jail for three days, and released with no compensation or support for having a smoke. My heart broke for this young man who was trying so hard to earn an honest living and had been subverted by an obvious bias against black men when it comes to incarceration. He had missed three days of work with no communication to his employer; would they even take him back? Having a cigarette could have cost him his job.

BREANNA

Working at a clinic where almost everyone lacks insurance, I see patients who are chronically unemployed or experiencing frequent episodes of unemployment between jobs. However, most of my patients are actually employed. This is not surprising as 74 percent of the uninsured have at least one full-time worker in the household and another 11 percent have a family member working part-time.[7] My patients clean houses, work for hotel chains, travel for construction jobs, and paint houses. Others work for restaurants, wearing their uniforms to their appointments because they will head into work as soon as we finish. I see the way my patients' employment status affects their health related decision-making. My patient with diabetes skips his appointment because work has been slow. He works hourly for a construction company and wasn't given many jobs over the last few months, so he can't afford even our low-cost office visit. Another patient tells me her low-cost mammogram will need to wait as she had to take a few days off work to care for her sick child.

My patients are industrious and seek work, even at the expense of their health. A few times a month I get asked by patients who cleans my house, hoping to pick up another job.

"I do," I laugh, "which means it's a mess!" The answer is my lighthearted way of avoiding the subject when in reality I am cringing. Even during their medical visit they are thinking about how they might get additional work. Cleaning one more house eases the financial strain a little bit.

Employment as a social determinant of health is difficult because health disparities don't simply exist between the employed and the unemployed. In the original Whitehall study, which identified a social gradient, the population studied consisted of *employed* British civil servants.[8] They differed in the type of work they performed. Those with lower-level jobs had worse health outcomes than those with higher-level administrative positions. Even though employment emerges as a determinant of health status apart from income and health insurance, not all types of employment bring health benefits.[9]

Employment is hierarchical in nature with certain positions or careers promoting higher social status. A higher social status is good for health. Those with lower occupational status (i.e., less education, lower pay) have poorer health.[10] Clearly, certain jobs can put people at risk for hazardous exposures, physical stress, or irregular sleep schedules. However, more significant is the role of job strain. Lower-level jobs are often characterized by high demands with little control in decision-making or task performance. Whether in manufacturing or service professions, the amount of work expected is ever increasing. At the same time, employees have little control over their work environment and few options for improving performance. This results in long-term health consequences.[11] Particularly, this type of work environment has been linked to cardiovascular disease, the number one killer in the United States.[12]

As a health care provider, this is particularly alarming. I know my patients are more likely to die from heart disease than any other chronic disease, so I check blood pressure and cholesterol and help them quit smoking. While helpful, perhaps I should also be asking about their work. Even when fully employed, parents are unable to lift their family out of poverty, increasing their risk for chronic diseases.[13] When compared to white-collar workers, blue-collar workers have higher rates of high blood pressure and their risk for hypertension actually increases the longer they are employed.[14] Chronic work stress doubles the risk of heart disease.[15] 'In other words, employment is a good thing, but in order to improve health, work environment matters just as much.

Barbara has been homeless for a little over a year. She talks proudly about having worked her entire life. She raised her son as a single mom and took care of several other family members over the years. She supported them while working full-time in the house she owned. The aging family members eventually died, and her son grew up and moved out. A little over a year ago she lost her job, and, as a woman in her late fifties with a host of health problems herself, she was unable to find another job quickly. Having spent her working years in the blue-collar sector and supporting various family members, she did not have a savings account that could sustain unemployment for a few months. As a result, she lost her house and moved into the city hoping for additional work opportunities. Since then she has been in and out of various housing and job-preparation programs. Barbara follows each program with the utmost motivation. She carries a manila folder with every certificate she's ever earned and a printed copy of every college credit she's amassed.

Barbara is also very willing to work. At one point she was given a job as a personal care attendant. She moved into a home where she was responsible for caring for three elderly women. While the job provided housing, it also required her to be onsite and available twenty-four hours a day. The house was far from public transportation and Barbara didn't own a car, so she stopped coming to her appointments and couldn't pick up her medications. One day she reappeared, telling me she was back in the shelter.

"I had to give them their medications each day," she told me regarding her personal care position. "But the medications didn't match their paperwork, and I couldn't find doctor's orders to verify I was giving them the correct medications." She went on to tell me she asked the owner about the situation. She snapped at her and told her to do her job.

"I got my certification as a medical assistant," she told me. "I know you can't just mix medications and give things without a doctor's orders." She told the owner she couldn't keep working without having the paperwork to verify what she was doing. The owner fired her.

"I guess it's better," she says, looking down at her hands. "It was stressful and I didn't have any way to leave the house or even get my own medication."

Barbara's situation is not uncommon. Even if she finds a job tomorrow, a near-minimum wage job won't immediately provide housing or transportation. She will still need access to health care and medications. While Barbara is motivated to work, she has multiple health conditions that make it impossible for her to stand for long periods of time. Her best option for escaping homelessness is likely through receiving disability. She would rather work, but as time passes her health is worsening and her options are dwindling. Her story is a microcosm for many across geographical locations and occupational settings. Rural communities watch industries leaving, and the hope of life-long financial security in long-term manufacturing jobs evaporate. Urban workers piece together multiple jobs with irregular schedules and little hope for advancement. Discouragement mounts and health status decreases, leading the poor to become sicker with few resources to get on a path to gainful employment.

CHAPTER 5

KOOL-AID IN A BABY BOTTLE

FOOD INSECURITY AND NUTRITION

VERONICA

I remember the first time one of our teenage neighbors, Latrice, came by our house to tell us she was pregnant. She was young and nervous, and we wanted to respond with grace and love. We congratulated her, and in an effort to encourage keeping the baby, we offered to throw a baby shower at our house in a few months. We reasoned that this was how Jesus would respond, choosing to celebrate new life no matter the circumstances.

The shower was wonderful! Our house had baby decorations everywhere, and we played what became a signature shower game in the neighborhood—candy bars melted in diapers that you had to smell and guess which brand of chocolate it was. There were diaper cakes, onesies hung on strings, finger foods, and gifts galore. There was laughing and singing and hugs and sometimes tears. It was always a beautiful event. In fact, it came to be something our neighbors looked forward to.

But I noticed something in myself. The more of these showers we threw, the more I felt an inner sadness and wondered if we were encouraging teenage pregnancy. While I firmly believe life is always something to celebrate, I was starting to notice a pattern that was not setting these new infants up for success. The cycle went something like this:

After announcing her pregnancy and being excited for a few weeks, a soon-to-be teenage mom would continue her normal lifestyle. Because of a total lack of prenatal counseling or care, everything from her life before carried over: poor eating habits, running around with the older guys, and limited sleep or creature comforts. They had smallish pregnancy bumps, even at the very end. Often, the first doctor they saw would be on the day they went into labor (via an ambulance to the Emergency Room)—usually delivering a very small baby prematurely. They would be released home without any in-depth knowledge about breastfeeding, caring for an infant, CPR training, or parenting. A few weeks later, when they emerged from their home to show off the baby, they would come over and I would see it: bright red Kool-Aid in the baby bottle.

While there are many issues that could be noted in this scenario, the most striking was always the Kool-Aid. To me it signaled what many studies have proven, that consumption of nutritious food is positively associated with their availability and affordability.[1] Red Kool-Aid was a reminder that healthier options were neither readily available nor affordable for my neighbors. As a result, from the earliest moments these infants were becoming addicted to sugar and not getting the necessary nutrition for proper growth and development. Because of a lack of nutrition for the moms, breastfeeding was nearly impossible, and a lack of money meant formula was prohibitively expensive. Certainly, teenage moms have access to WIC (federal assistance program for women, infants, and children), formula vouchers, and food assistance, but they lack access to health education on the importance of milk, breastfeeding, and proper nutrition to develop their baby's brain. On more than one late night I scooped two cups of sugar into a plastic bag and gave it to a new mom, knowing full well they were going home to make red Kool-Aid for their newborn. In hindsight, I should have kept baby formula on hand for moments like these. At the time I reasoned that they had no other options to soothe a hungry baby, and Kool-Aid was better than nothing.

Food insecurity is a serious issue in our country, plaguing forty-two million people in the United States, thirteen million of which are children.[2] Our former neighborhood was a "food desert," a designation indicating that healthy food options are more than three miles away from the community. This might not sound like a big deal to someone with a car and full tank of gas, but for families without transportation this means their "grocery store" is the local gas station. As a result, kids grow up drinking a lot of soda and eating candy, chips, and beef jerky. Every time a new kid would come to our house, they would see the fruit basket in the kitchen and insist it was fake until they took a bite out of the very real fruit. For some it was the first apple or pear they had tasted in a long, long time. The first time our neighbor Isaiah saw our spice rack, he asked us if it was weed. To this day I'm not sure if he believed us that it was oregano and basil. In addition to issues of access, these kids are bombarded with advertisements, which often form the basis of their knowledge about food. One study found that black youth viewed 70 percent more food-related TV advertising than their white peers and almost twice as many TV ads for sugary snacks and beverages. Some major food brands disproportionally target their TV advertising to black or Hispanic consumers, which ultimately leads to poor health.[3]

The issues of food quality and food quantity were both present at the same time. At best, the food quality was poor, and if kids had food, it was just unhealthy. At worst, food quantity was the issue, and I would see kids squirreling food away in their mouths. Squirreling was a way of hiding food like corn starch in your cheeks so that you could save some for later. This did little for kids' nutrition and had disastrous, long-term effects on their dental health. In fact, another charitable dental clinic on Atlanta's Westside confided in me that squirreling was the hardest issue they had to overcome with their periodontal patients.

SNAP (the Supplemental Nutrition Assistance Program, formerly known as food stamps) alone did not solve these issues for two reasons. First, families were often pushed to make desperate decisions. More than once Brandon's family sold their food stamps to pay for routine shots for their infant. Food longevity was also a concern. Brandon's family lived in a house with multiple families under one roof, and they would steal from each other. Besides this point, they did not have a refrigerator or a means of preserving fresh foods. Prepackaged ramen noodles or chips were easier to store and hide.

When we did our weekly grocery shopping, which required driving out of the neighborhood to a decent grocery store, we would take the kids with us. They were always amazed by all the cars on the highway (aka Atlanta traffic) and enjoyed the adventure. For most it was the first time in their life they had left the neighborhood. Christopher Burke, director of community and government relations at Georgia Tech, grew up in a working-class neighborhood and understands this dynamic on a personal level.[4] Burke refers to it as "invisible borders and barriers" that keep people from leaving their neighborhoods because they don't see themselves any place else. He says even if they had viable transportation, they might not leave because they haven't had any real exposure to a different environment. So we did our best to take the kids places, and at the grocery store each kid would get a "budget" and be challenged to come up with a meal plan. Without realizing it, they were doing math and learning some of the basics of measurements and recipe creation.

This practice also encouraged the consumption of new fruits and vegetables. I remember Isaiah trying asparagus for the first time. He had insisted it was "disgusting," "gross," and would make him throw up. When we got to the bottom of it, he had never tasted it before and just thought it looked weird. With a little salt and a touch of ranch dressing, it became his favorite food. I remember with fondness how many times Isaiah walked up to the shopping cart with a smile on his face and asparagus in his hands. I'm not going to lie; I don't even like asparagus like that!

In Isaiah's case, his mother didn't grocery shop because she didn't have the transportation. Other parents did make it to a grocery store, often a Kroger, but still struggled with food quality. The perception was that the items marked "10 for 10" were a deal (often prepackaged junk food) and healthy items (like fresh fruits and vegetables) were "too high" (or too expensive). It was common for families to spend their food stamps on one amazing meal at the beginning of the month and then eat poorly until the next check. One time I asked Isaiah's mom why they did this instead of spreading the money out so they could eat all month. She said, "I spend so much time failing to provide for my kids that when I can, I treat them to something special."

There were some efforts to address this situation. The local food bank had generously loaned a sizable plot of land to the community for a neighborhood garden. My neighbor Randy, a master gardener, was very passionate about this project and believed that above all else if people knew how to grow their own food, they would never go hungry. The garden was a great idea, but it died on the hill of logistics. It was a long walk from our house, and for many years we had to bring our own water. Walking a mile each way in the heat of summer carrying three one-gallon jugs of water was a tough sell. The kids lost interest, and over time the garden became overgrown.

Years later I met an African American couple at a local food conference who had revived the garden project. Rather than running a volunteer-based community garden, these folks were farmers who had moved in from the suburbs and made it their full-time occupation to run an organic, low-cost garden for the community. The wife confided that she often feared for her safety and was disturbed by the constant gang-related violence. But then she paused and with great seriousness said, "Gang members need to eat too."

With all these issues surrounding food, it's no wonder that sometimes children are referred to as "more mouths to feed." Keeping kids warm, clothed, fed, and cared for is no small feat when you're living in poverty, and the current governmental assistance provided to poor Americans is too little. Certainly programs such as SNAP and Medicaid help, but even still, infants born to mothers with less than twelve years of education are twice as likely to die during their first year of life when compared to mothers with sixteen or more years of education.[5] In my observation of my neighbors, a lack of education and support for parenting often led to hard decisions, not what is sometimes assumed to be a nonchalant or callous attitude toward their children. Some of the most enthusiastic, supportive lovers of children I have ever known were my former neighbors. When I became pregnant for the first time, it unfortunately ended early in a miscarriage. I had just received the news and pulled into my driveway when Fran, a dear grandmother who lived across the street, came running over hollering "how's my baby?" with a huge smile on her face. She was the first person I told about the news, and with huge tears falling from her own eyes she hugged me and we stood in the front yard and wept together. I could feel her devastation matched or exceeded my own for the loss. A few years later when I became pregnant again with my daughter Aubrey, everyone we knew in the whole neighborhood congratulated us. Brandon, with great sincerity, said, "I'm really happy for you guys. You've spent so many years loving on other people's kids; I just know you're going to love being parents."

We *have* loved being parents. Partly because it's amazing and wonderful, and partly because we have the resources and education to cover our family's basic needs. But what happens when family economics get squeezed in the context of poverty? Unfortunately, heartbreaking decisions can ensue.

One notorious and tragic story known across the neighborhood happened after our teenage neighbor Tilly gave birth to her second baby. Tilly was abused and neglected by her own mom and had moved out of the house as a very young teenager, living with one boyfriend after another. Tilly was thin, very shy, and effectively homeless. When someone approached her and offered her two months' rent in cash for her second born, she sold her baby boy. She reasoned that the money would take care of her small family for a while, so she wouldn't have to worry about where the next meal was going to come from. When you feel insecure about the most basic determinant—food—everything else is negotiable.

BREANNA

For our neighbors and patients facing food insecurity, their supply of food is not adequate to lead a healthy lifestyle. I regularly see patients who will warn me that their blood work will not be good at their present appointment:

"I didn't have much work this month, so I wasn't able to eat as healthy as I had planned."

"I was able to get a construction job out-of-state, so I mostly had to eat out."

"We only have one option for each meal at the shelter, and there is often just pasta and bread."

These aren't excuses. These statements represent the realities my patients are facing. They often feel unable to make healthy choices due to financial constrictions. The hallmark of chronic disease management for conditions like diabetes and high blood pressure is a dietary change. But what do you do when that is not an option? For many of my patients it is easier to take more pills than find an affordable, accessible way to change their diet. Diets high in simple carbohydrates and sugars are generally cheap and widely accessible. Fresh produce requires access to a grocery store versus a corner market. The use of fresh foods requires planning, correct storage, and preparation. Opening a bag of chips or driving through a fast food chain presents none of these barriers. When I talk to my patients about healthy eating, I am often starting with the question, "Where can you find healthy food?"

Even when healthier options are available, the knowledge of how to incorporate them into a balanced diet can be elusive. The longer I have worked, the more I have determined that most people don't have an accurate concept of what it means to eat healthily. Robert is an example of knowledge as a barrier to adequate nutrition. Looming at just under four hundred pounds, Robert is as kind as his stature is large. He works in the music industry, which provides a good living, but income is inconsistent and without benefits. He allocates money each month to afford the growing list of medications he needs to manage his blood pressure and blood sugar. As a single man in his late thirties, he readily admits that a drive-through provides at least one of his meals each day and sometimes more.

When I first told him he had prediabetes, we talked about simple carbohydrates. He agreeably listened, but his lifestyle made it difficult to implement the changes. While he could afford healthier food choices, the planning and preparation required were complicated for Robert. He got busy with work and missed a couple of appointments. One day, he was feeling bad and decided to check his sugar. To his shock it was over 400, and he immediately came to the clinic where we started insulin. I could tell he was overwhelmed, so I gently brought up the issue of his diet again. I asked if he might consider starting by cutting out some of the fast food. Realizing this was the staple of his diet I suggested he start with eliminating french fries and soda entirely. The expression on his face confirmed this would be a major change. He came back two weeks later defeated.

"I stopped fast food almost completely, and I gave up soda!" He exclaimed. "I started drinking only juice, and I've been eating fruit but my sugars aren't improving much." I paused for a moment and congratulated him on his efforts to change. Then I went on to explain that juice has as much sugar as soda, and, while healthier than a candy bar, fruit contains sugar as well. He had earnestly tried to make a change but had no concept of carbohydrates and sugars.

Many of my patients are trying to make healthy decisions but lack the knowledge to do so. Teaching takes time. Robert and I continue to work together, and he has made huge improvements. Each time we meet, I focus on one new concept or tip related to his diet, and he sees the clinic nutritionist as well. However, he spent the first nearly four decades of his life learning about food from his family, advertisements, and community culture. Unlearning and relearning is difficult and takes time.

People of lower socioeconomic status are less likely to obtain higher levels of education and less likely to have health insurance. Not surprisingly, people with more education are more likely to learn about health than those with less education.[6] I regularly see people like Robert who are making poor health decisions. However, I find that often this stems from a lack of knowledge and exposure to healthy lifestyles. Simply telling Robert once that juice and fruit won't help his diabetes doesn't counterbalance the understanding of health he has built from observation and public messaging. The chronic stress experienced by people of lower socioeconomic status also works against them. Chronic stress signals the body to accumulate abdominal fat around the waistline, which worsens many chronic health problems, including diabetes and heart disease.[7] Most of my patients aren't seeking to make poor choices. They are doing their best with the access and education available to them.

As a health care provider, I often wonder why someone has to be diagnosed with a condition like diabetes in order to receive education on nutrition. When people lack health insurance, they are usually not seeing a health care provider until there is a problem. If

I am the first person to tell them about nutrition, it is already late. However, health education is often best accomplished outside of the clinic and closer to home. Adjacent to the clinic is Good Sam's one-acre urban farm. Throughout the spring, summer, and fall, the farm team sets up a daily farmer's market in front of the clinic. Patients and community members stop on their way to and from appointments to look at the fresh produce being sold at affordable prices. On occasion I've walked by the market and heard someone inquire about a vegetable they haven't seen before. Our farm team fields questions and enthusiastically shares their favorite way to prepare the vegetables. Pretty soon everyone is discussing favorite recipes and sharing tips on using produce. The conversation is organic and social without the graphs, goals, and logs we are using inside the clinic. The market is the perfect intersection of food access and education. It is a small step toward improving nutrition in the community but, like any garden or farm, it can grow.

CHAPTER 6

LONGEVITY AND LEARNING

EDUCATION AND CHILD DEVELOPMENT

VERONICA

Caleb came over after school most days, so when I heard the knock at the door I didn't expect anything out of the ordinary. However, when I opened the door I saw it—a large welt right above the right side of his lip. It was bleeding and looked like it had not been cleaned.

"Oh my gosh, what happened?"

He shrugged shyly and just said, "I was stabbed in the face with a pencil at school."

"What?" I roared. Caleb was like a son to me, and I was suddenly ready to hurt someone.

He proceeded to explain that he had gotten into an argument in class with someone who was making fun of him. Since we had started tutoring Caleb a couple years prior, he struggled with classmates poking fun at his newfound interest in school. He described it as being "white" to do well in school, and kids often picked fights with him for caring about his grades. This racial stereotype likely stems from education disparities seen throughout the United States. A study done in similar Detroit neighborhoods found that African Americans in high-poverty areas had overall lower education levels and lower income than whites.[1] Caleb was caught in a constant tension between the negative culture of his peers, which discouraged learning, and his own desire to graduate from high school and go to college. Normally he walked this line well by adjusting his approach when speaking to different groups of people, but every once in a while conflict erupted.

The reality is that at Caleb's school a prevalent fight culture existed and almost demanded everyone's participation. Fighting was how things were resolved, and students who didn't fight put themselves and their families at risk for worse consequences. Peer pressure largely discouraged academic achievement and encouraged crime, fights, and substance abuse.[2] Caleb told us the school police often had to resort to forceful measures to break up large group fights. Classroom teachers were often scared or beyond the point of fatigue, such that on the day Caleb was stabbed, his teacher merely reprimanded him and the other young man and kept moving on with the lesson.

I knew Caleb's mom couldn't do anything about it. Not because she's an uncaring mother, but because she was severely mentally ill and rarely left the house. She also relied heavily on Caleb for survival. Caleb had a kind heart, and despite incredible neglect he loved his mom and took good care of her—giving her his hard-earned money to buy illegal drugs or groceries. This is not uncommon. Boys often take on the role of men in their homes and carry the burden of providing and protecting at an early age. At one point Caleb had the opportunity to attend a local charter school free of charge if his mother would simply sign the papers and agree to do some minimal volunteering on campus. Although Caleb's grandmother understood the value of this opportunity and pushed hard, his mother declined. Unfortunately, when it came to parental expectations, there was a different value system about education that was reinforced by the local, largely segregated school. Segregated schools generally underperform on standardized testing, have less-qualified teachers, less access to serious academic counseling, and higher dropout rates when compared to schools in middle-class areas.[3]

Other barriers to learning existed for Caleb. He often skipped school to care for his mom, or he was up late playing video games with older cousins and no one bothered to wake him up the next morning. At a very young age he was responsible for understanding the value of education and disciplining himself to get up and ready for school. On more than one occasion Eric or I drove Caleb and his siblings to school after missing the bus. Caleb also attended a school that required uniforms, and he barely had one full set of clothes. His family didn't own a washing machine and rarely went to the laundromat. Lack of access to a shower or the opportunity to wash his clothes kept Caleb home from school and caused him great shame.

I am convinced that sometimes the only thing that got Caleb to school was knowing that he would be fed, or the fact that classrooms were heated in the winter when his home was not. When schools closed in Atlanta the year we had a snow and ice storm that shut down the city, Eric and I worried the kids would get frostbite or hypothermia. So, we offered Caleb the choice between some cash or a ride to Target to purchase gloves, hats, sleeping bags, and blankets. Caleb picked the latter, only to get in major trouble with his mom for spending money that way rather than give her cash outright.

We poured more time, love, attention, and resources into Caleb's life than most education-based nonprofits can offer. Because we lived across the street, we were always available to give food, school supplies, encouragement, tutoring, mentoring, and the like. We toured colleges, rewarded improvements in report cards, played sports, went on educational field trips, and lived out the very essence of life-on-life discipleship. We exposed Caleb to a lot, but Christopher Burke explains why some exposure is not enough:

> The flip side to exposure is that if it stops there, it doesn't create a pathway forward. Robotics workshops are not enough. You don't get into Georgia Tech by doing a little coding at a

> community center; you get in by learning advanced math and critical thinking. Passing middle school algebra is the biggest factor regarding whether a student can grow up and do computational thinking and be successful at advanced math—which is what entrance to Tech requires.

In addition, schools in low-income neighborhoods often do not have the resources to offer algebra or the foundational math necessary to go into computer science.

Caleb dreamed about going to Georgia Tech. We walked the campus, and he would point out his favorite things about the school. But in the end, Caleb dropped out two years before graduating from high school and started working at a local warehouse doing manual labor—the barriers to learning were insurmountable. And still today, just above his right lip, an eraser-sized scar remains.

As a sixteen-year-old high school dropout, Caleb faces a lifetime of poorer health and increased mortality. Statistically, he will likely die earlier than those who graduate from college and will be more likely to die from a chronic disease.[4] Further, these issues disproportionally impact lower-income minority youth, so the same would ultimately be said for many of the other young men in the neighborhood. Despite interventions in the American South, a child living in Atlanta only has a 5 percent chance of rising out of poverty.[5] The American dream is fading.

BREANNA

The resources Veronica and Eric poured into Caleb's education may seem extreme, but the importance of education on health status and life expectancy cannot be overstated. Adults without a high school education die on average nine years earlier than adults who graduate from college.[6] The numbers are even more dramatic when considering race. Caleb didn't finish high school, making his predicted life expectancy 14.2 years less than a white man with sixteen years or more of education.[7] The mechanisms for this are many. Education is associated with better jobs, higher income, and improvements in social status. In the United States, health insurance and health care access are also closely tied to employment. Over a fourth of adults who did not complete high school report being unable to see a doctor due to cost.[8] In addition, those with higher levels of education have better access to information about healthy living and are more likely to choose healthier behaviors. For example, 24 percent of people without a high school degree smoke compared to 7 percent of people with a bachelor's degree.[9]

As a result of lower social status, lower income, and poorer health behaviors, people with less education are more likely to have heart disease and diabetes.[10] The more education a person receives, the less likely they are to suffer from heart disease, high blood pressure, and stroke.[11]

The relationship between education and health is particularly troubling given the narrow window of time in which to reverse the damaging health impact of low educational attainment. Caleb's education trajectory started long before Veronica and Eric became involved in his life. My husband, Matt, is an early elementary school teacher. When we first moved to Atlanta, he taught first grade, right around the corner from the clinic where I started my career. Having completed his student training in the rural Midwest, teaching at an urban Atlanta school where nearly 100 percent of children qualified for free lunch was a bit of a culture shock.

I will always remember his second year of teaching as "the year of Treveon." Treveon was a teacher's nightmare. He screamed and threw fits, he yelled profanities and said things no six-year-old should have ever heard; he threw books, desks, chairs, and anything else that wasn't bolted to the ground. He was also a very low-performing student academically. He couldn't read or keep up with the other students, likely a result of his constant fidgeting and inability to concentrate. Matt, along with the school administration, intervened to a degree. Matt tried increased compassion and affection, discipline and stricter boundaries, and creative seating arrangements. Nothing worked. The school implemented an intervention plan, and Treveon received a psychological evaluation. He had a laundry list of diagnoses ranging from learning disabilities to ADHD. He was given medication to help, and it was effective but not sufficient to fix the problem.

To understand Treveon's troubled school performance requires insight into his life outside of school. Treveon was raised by his grandmother as his mother worked and lived on the streets. He knew his mother and would see her out on the streets occasionally, but she wanted nothing to do with him. If Treveon's father was known, he was never in the picture. His grandmother was overwhelmed and couldn't keep up with his needs, appointments, and prescriptions. When he did have the medication, she would wait and give it to him after school to make his time with her more manageable. At the end of first grade, Treveon still couldn't read.

In 2016, Matt got a text from an old colleague telling him Treveon was dead. He had stolen a car with a friend and fled the police in the stolen vehicle. The police chase ended in a fatal crash. Treveon was fourteen. I could see Matt's pain in receiving the news. While his career has been filled difficult and vulnerable students, Treveon represented a particular failure in the systems designed to protect vulnerable kids. The educational, social, and medical resources came too little and too late. Treveon's tragic fate seemed already in motion by the end of first grade. Inability to protect vulnerable kids translates into unhealthy young adults who are less likely to succeed in school and find good jobs later in life.[12]

Treveon's story is a reminder of the importance of intervention and resources early in life. By first grade, at the age of six, Treveon was already behind and frustrated in school. Learning to read in first grade is the start of future academic attainment that has significant implications on adult health status. By third grade, students transition from learning to read to reading to learn, meaning that an inability to read hinders learning across all subjects.[13] A study in the Chicago Public School system found that 80 percent of children with above-average reading scores in third grade graduated high school compared to 45 percent of those with below-average reading levels.[14] In addition, less than 20 percent of third grade students reading below grade level went on to attend college compared with 60 percent of those reading above grade level. Similarly, a study in Georgia found students who did not meet expectations for reading levels in third grade were less likely to graduate high school and less likely to take the SAT or ACT.[15] Waiting until high school to intervene is too late.

The knowledge discrepancies Matt sees in first grade start before any child enters the school system. By the time kids enter kindergarten, children from disadvantaged backgrounds have lower verbal and cognitive abilities when compared to their peers of higher socioeconomic backgrounds. In fact, at two years old, children of lower socioeconomic status (SES) are performing at the same level as higher SES children six months younger.[16] Children's early vocabulary is learned from their parents. The famous study by Betty Hart and Todd Riley found that children in families on welfare heard thirty million fewer words than children in working-class families by the age of three.[17] This gap correlated with poorer school performance by third grade. Children with lower literacy grow into adults with poor literacy skills, fewer job options, and lower earning potential.[18] The reasons for this word gap are many, and researchers are continuing to determine the factors contributing to this gap. Parents faced with the stress of poverty, multiple jobs, and transient employment lack the resources and time that promote vocabulary acquisition in children regardless of their desire to do so. These parents may also be limited by their own literacy skills; adults with the lowest literacy skills are less likely to read to their children or engage in literacy promoting play.[19]

Listening to Matt's experiences over the years has convinced me of the importance of early childhood education and interventional programs. There is still much work required to create equitable educational systems. Yet, despite his frustrations and setbacks, Matt is heading in to his second decade as a classroom teacher. He has as many humorous stories as somber ones. One cold day he had a line of six and seven year olds who couldn't figure out their jacket zippers in order to go to recess.

"Why can't any of you zip your coats?" he teased them.

Ziarah cocked her head and plainly told him, "Mr. Lathrop, we only just got born!"

Six year olds are still very much babies, and their educational journeys don't have to end in high school. Investing and intervening early can add years of healthy living to the life expectancies of the little people in Matt's class.

CHAPTER 7

WHEN HOUSING HURTS

ENVIRONMENTAL FACTORS

VERONICA

I grew up in the comfortable suburbs of central Florida, which means I had no idea what it was like to be surrounded by poor social determinants of health. I had rarely observed any housing or neighborhood blight, much less witnessed the compounding and hopeless effect of poverty and crime on a community over generations. It's a different world. In an analysis of 171 of the largest cities in the United States, it was found that the worst urban context in which whites reside is better than the average context of black communities.[1]

The moment I exit the highway, driving to Southwest Atlanta is noticeably different from driving to the suburbs. The first stoplight is a popular spot for homeless folks to camp out with their cardboard signs. The most common gentleman I'd see was a thin, older man who was always drunk, holding a sign that read, "Why lie? Need a 40 oz." On the side of the road where he stood, empty beer cans and trash littered the dark, stained curb.

Turning left onto the main street I would drive over a short bridge that always had police cars parked on it, most often with lights and sirens blaring. Driving a little further I pass a Popeye's, Church's Chicken, Taco Bell, gas station, Chinese restaurant, and a sub joint named Gut Busters. A McDonald's, Krispy Kreme, and a soul food restaurant were located just off the main drag. This is the extent of the food options, and even if I were in the mood for fast food I wouldn't have stopped—police cars and active arrests were often taking place.

There's a large intersection that is always tricky to drive through because of constant speeding and jaywalking. The turn signal is short, so folks always run the light. I remember one day driving home there had been a shooting and a policeman was still in the middle of the intersection holding a gun in the air and waving me through. With my infant daughter in the back and my aunt in the passenger seat, I drove through the intersection hunched way down looking up over the steering wheel just high enough to see the road.

After clearing this intersection I pass a junky pawn shop, check cashing store, the local MARTA station, a few more run-down gas stations and a huge unoccupied warehouse completely surrounded by barbed wire. The sidewalks are totally broken up, and a few historic but crumbling old buildings line the street. Finally, I pass the open-air flea market, one of the neighborhood's major economies, and a big white building that always had fancy black convertibles parked outside at all hours of the day and night. To this day I do not know what business they were running out of this space; the cops never went there.

Driving down my street there was tons of kudzu overgrowth that engulfed vacant homes and was making its way up the light poles. All the homes were in disrepair, including the occupied ones, with shattered windows and trash everywhere. Empty liquor bottles and graffiti were everywhere, and the smell of trash or weed was in the air. Everything was literally broken.

Usually, a loud block party was going on, and the ATV gang undoubtedly would be revving up and down the street, popping wheelies, and driving at dangerously high speeds. Most of the riders drove in from outside the neighborhood because they knew they could get away with driving dangerously. Our community was often a destination for bad behavior because local law enforcement would look the other way. For all the times we called the police to complain, the madness only continued and the gang grew bigger and drove more frequently. I noticed early on how differently the cops responded to issues in the Southwest Atlanta compared to my suburban upbringing. Whether it was because the local police precinct was overburdened with calls or under resourced, they were routinely sluggish to respond.

When first responders did come into the neighborhood, they would often be present to transport a sick child or elderly adult to the emergency room for treatment. In my neighborhood there was no clinic. Families would wait until a medical emergency or a mother was in labor and call an ambulance. In the winter an occasional fire truck usually meant someone had built a front porch fire for warmth that had grown out of control.

When we pulled up to our home, the neighborhood kids would immediately come running over, often in the middle of a weekday. For some reason, either their own choice or their parent's neglect, they were skipping school. They'd be snacking on chips and coming over to ask for a piece of fruit or a meal. If there was a new child in the group we would ask them, "Where do you stay?" We learned early on that the sense of impermanence in housing penetrated to the language level. No one "lived" anywhere; they were just "staying" temporarily.

Many days the kids skipped pleasantries and instead of "Hi" it was just, "Can I have a banana?" Occasionally they didn't ask for anything and just wanted to get out of their home because it was crowded and they "needed some space to think." Many wanted to know if the "game room was open today" because they were bored and wanted some attention. Or they just wanted to play with the dogs or sit on the porch and chat about life, school, or family issues. Even though there was a park less than a mile away, the kids opted not to go because it was a known drug dealer hangout and considered unsafe.

Young peoples' health is compromised when they fear walking outdoors and instead sit in blemished buildings in need of repair with no access to nutritious meals or physical exercise.[2]

On one overcast day, Eric and I were walking with our toddler in a stroller through this park past the playground. We looked over to see one kid trying to stab another with a pair of scissors and a teenager surrounded by five little kids passing his joint around. Ironically this was all happening right outside a recreation center, which, at that very moment, was running educational programming for kids. Having spent so much time pouring our lives into kids in the neighborhood, this moment was particularly devastating. Further, the park was supposed to be a safe place for kids to be kids, but instead ours was discouraging safe exercise at best or encouraging drug use at worst.

As challenging and heart-wrenching as these environmental factors were, local violent crime by far made the biggest impact on my family's health. It was pervasive and constant, interrupting even the most benign moments of life—like working in the garden. It wasn't just my neighborhood. Other communities of color and low-income communities are plagued by high crime rates, insufficient services, and other harmful attributes that compromise safety and health.

PLANTING TULIPS

One day I was hunched over a small red shovel, sweating profusely from battling the mesh laid down years before to smother weeds. I was literally cursing myself now for this decision, as it made turning over the soil to plant my tulip bulbs a full body workout. It didn't help that at the time I hated gardening but could not convince a landscaper to come to the Southwest side of Atlanta to maintain my small front yard. This could be explained by high crime rates influencing the quality and availability of services and economic opportunities for businesses, which affects their willingness to locate in the neighborhood.[3] On more than one occasion I had arranged for a quote only to get a call right before the appointment with a cancellation and the excuse, "I just realized you are out of our service area." So with great frustration, I was doing it myself. *We would have tulips!*

Just then an unfamiliar man walked up wearing a wooden board around his neck that read "Merry Christmas" (it was November). I was tickled and said, "Hey, I like your sign." He thanked me and complimented my yard before asking if I'd heard about the man who was shot and killed on the corner this past weekend. When I say "the corner," I mean two houses down from mine at the next intersection. He seemed spooked by it and was looking for a soul to commiserate with. I told him I hadn't heard about it, to which he exclaimed, "You didn't hear the gun shots?" I usually do hear gun shots, the domestic abuse playing out on the streets, and the police sirens at all hours of the night, but somehow I missed this. So I mumbled, "No, I didn't." At that he turned and kept walking, leaving me to plant my now very silly and meaningless tulips. In a neighborhood ripe with violent crime, what good are tulips?

Before this moment I had hypothesized that the best way I could help my neighborhood was to make my own home a place of beauty. I reasoned that if I cared tenderly for my house, it would serve as an inspiration to my neighbors and encourage others to move into the neighborhood, occupying vacant homes and contributing to the community. Working on my house had also become a source of positive therapy. Every time I was blazing mad about the conditions of the neighborhood and its adverse effects on residents, I would pour that energy into a home-improvement project. As a result, after a few years my one hundred-year-old house looked pretty darned good.

But having a nice-looking house hadn't inspired a thing. All the houses around me were still vacant or filled with renters who didn't care about the aesthetics of their temporary residences, or owned by landlords who were making a buck but not maintaining their properties. But here I was because I didn't have any other real skills I could give my neighborhood, trying to force my thumbs to turn green. I had even flirted with the idea of adding a bird house and a butterfly garden. I wanted people to walk by my house and, even if they were carrying heavy burdens, look up and for just a moment have a sip of loveliness. I wanted the flowers to remind my neighbors that beauty still exists in nature and belongs in their community.

I was slapped back into reality by the lost life of a man I hadn't had the pleasure of meeting or even knowing he had died. In fact, he passed away like it never even happened. I checked the news for weeks afterward, and although I found scores of articles about other men dying violently in Southwest Atlanta, he never made the press. I learned quickly that Googling our neighborhood zip code mostly resulted in articles about murder and arrests, so I eventually started a blog to share positive news about the neighborhood. For a long time it was the only internet search result for our community that didn't refer to violence.

Crime was prevalent because it was easy to get away with. We called 911 countless times in our early years in the neighborhood. Even after the city of Atlanta implemented a fee for "false alarm" calls, we continued, despite the fact that it was rarely effective; the police always arrived too late. Many low-income neighborhoods lack fire stations and police monitoring the streets.[4] The only "person" who ever caught an intruder on our property was our ninety-pound male Akita named Po. A man on drugs had crashed his car into a light post near our home and took off running. High as a kite, he ran straight through our fence only to come face to face with Po, who chased him to the front yard and held him by a pant leg, unharmed, until the police arrived. Po was regarded as a hero by the kids in the neighborhood from then on.

But on this gardening day, I was only half finished with my planting, and given this news I seriously contemplated throwing the rest of the bulbs away. But before I could do that one of my neighbors, a young man named Carl, paid me a visit. He helped me back away from the cliff of giving up—a cliff I was visiting more frequently these days. We chatted about music, books, art, family, and gardening. He suggested adding a butterfly garden, and I wanted to hug him as I exclaimed, "I've been thinking about that very thing!" Carl was always a mystery. He was a sweet, thin, contemplative teenager with thick, wire rimmed glasses who asked thoughtful questions and loved Katy Perry's song "Firework." Whenever it came on, he sang and danced to it like it was his life anthem, telling his story and insisting to a harsh world that he was going to be somebody. When I was near the end of my pregnancy with my first daughter, he brought me flowers plucked from someone's yard and said, "Remind me when that sweet cherub is going to be born." Every time he came by we would talk and smile and laugh, and he would respond like a young flower arching toward the sun.

Carl's presence was a simple comfort on that November afternoon, and so I finished the job. I didn't just finish it for me but for the unnamed man. It's a strange economy to respond to murder with tulips, but it's all I knew how to do. I dedicated them to him in my heart and promised that when they burst forth in the spring, I would honor his memory.

Hearing this kind of news was so common; I eventually became numb to the realities of violence and reacted by drinking too much. I now know how and why a young woman can be sitting on her front porch at 2 p.m. in the afternoon nursing a glass of wine. My heart ached so badly. I was afraid of the violence and felt impotent to do anything to stop it, reacting to this sense of powerlessness in a self-destructive way. I do not believe this behavior emerged from "making bad choices" alone; rather, I was reacting to a frightening environment and trying my best to cope. I believe I and many of my neighbors were doing just that—living as best they could in a harsh environment.

So I would sit on the porch having a low-level panic attack, stress coursing through my veins, staring blankly at the abandoned house across the street, and gulping a goblet of merlot. My face was totally broken out in stress acne, and I would turn seemingly insolvable safety, financial, and life issues over and over in my mind—never landing on a satisfying solution. Instead, I'd pour myself another glass of wine, sit back, and think about how gravely I underestimated the impact of environment on health status.

BREANNA

From the high-rise in downtown Atlanta that houses the offices of the Atlanta Volunteer Lawyers Foundation, Cole Thaler can almost see Centennial Olympic Park.[5] The park is a reminder of the 1996 Summer Olympic Games as well as the end of public housing in Atlanta. The projects have been long demolished, but the housing crisis is still very much a reality. "Atlanta is very thin on housing resources," he tells me. Thaler is the Safe and Stable Homes project director for the Atlanta Volunteer Lawyers Foundation (AVLF). The organization's mission is to "create safe and stable homes and families by inspiring attorneys to fight for equal justice."[6] In particular, Thaler helps clients obtain justice against unlawful landlords and relief from deplorable housing situations.

Unfortunately, unsuitable living conditions are not rare in the United States' affordable-housing crisis. The Department of Housing and Urban Development recommends that people should not pay more than 30 percent of their income for rent or mortgage, however two million renter and homeowner households are currently paying more than 50 percent of their income for housing.[7] There are not enough affordable rental units for those who need them. Extremely low income (ELI) families are those whose incomes do not exceed the higher of the federal poverty level or 30 percent of the area median income, as defined by the Department of Housing and

Urban Development (HUD).[8] There are only twenty-eight adequate affordable rental units for every one hundred ELI rental households and no US county has enough affordable housing for all ELI renters.[9] Lack of affordable housing is the most important cause of homelessness in the US.[10]

In the counties that house the city of Atlanta, there are forty-six affordable, adequate, and available units per one hundred ELI renters in Fulton County and twenty-four such units per one hundred ELI renters in DeKalb County.[11] By 2011, Atlanta became the first US city to eliminate all project-based public housing.[12] The end of public housing did not correlate with an improvement for low-income tenants. Not only are affordable, adequate units lacking, closing the projects also eliminated supportive communities. Thaler explains that the motivation for shutting down the projects was the coming of the Atlanta Summer Olympic Games and concerns about how the city would look to outsiders.

The city didn't recognize that the projects were happy homes and there was a sense of community. "Residents recognized that crime had become a problem," Thaler explains. "They wanted the crime to be addressed, but a lot of people didn't want to move." Treanda (Tre), my friend and colleague who grew up in the Techwood Housing Projects from 1976 to 1983, has fond memories of her childhood there. "I never felt poor," she explains. "There was a strong motherhood presence. Lots of moms there banded together. We always had supervision and meals."[13] Her sentiment is shared by others. Griff Tester and his associates argue that many residents experienced a sense of loss, leaving their community in the process of relocation.[14] Relocation did result in the majority of residents moving to areas of less poverty and crime, yet nearly a third described the process was "hard, terrible, or traumatic with stress."[15]

The best approach to unhealthy neighborhoods is often not tearing down structures and moving people out. As illustrated by the closing of Atlanta's projects and Tre's experience, disrupting social bonds has consequences. As Otis Rolley, former director of planning for Baltimore, explains, a city is made up of social networks.[16] He argues that these social networks that exist even in the most impoverished areas of a city should be valued and invested in throughout planning and development activities. While substandard living conditions are problematic for health, the elimination of the protective relationships between neighbors can be equally devastating.

Thaler sees cases of substandard and dangerous living situations every day. In particular he remembers a woman in her fifties who had relocated to Atlanta from New Jersey to be closer to social support. She was living off disability benefits and her housing options were limited. She moved into an apartment with severe electrical problems. She would regularly lose electricity, and the landlord would send out a "useless maintenance guy." The parts he ordered never seemed to arrive and the problems kept mounting. Lack of electricity is problematic for everyone, but particularly for this woman because she had sleep apnea. Her life depended on a CPAP machine to deliver oxygen to her every night. Eventually, the power bill reached $700. "She had to choose between paying rent or risk of death in her sleep," explains Thaler. She chose the power bill.

"She did nothing wrong," recounts Thaler. "She did what any one of us would have done. She was completely at the mercy of her landlord to make the repair. This is an example of how we set people up to fail." The landlord filed an eviction and AVLF took on her case, filing a counterclaim, which prevailed. The rent was waived and she was able to move out.

The outcome could have been very different, especially if she had lacked representation. Eviction means not only acute loss of housing but creates a record affecting ability to get housing in the future. Thaler has also seen this scenario. While at court fighting her eviction, Thaler and his client watched a young woman with children facing eviction before the judge. "She was alone," describes Thaler, "and didn't have a good case." The judge ruled against her, and she was given seven days to move out. "She was crying and trying to wrap her head around what just happened," remembers Thaler. As she was walking out, Thaler's client jumped up and handed her a number where she might be able to find a rental despite the eviction notice. As she sat back down she said, "Maybe that's why I am here today, to help her." Maybe the young woman's case would have turned out differently had she had representation. Regardless, she stood with her children with only seven days separating her family from potential homelessness.

Living conditions can clearly influence health through chronic stress, as illustrated in Veronica's experience. Poor living conditions can also have a more immediate effect on health status. I have written countless letters to landlords recommending mold inspections for patients with uncontrolled asthma. I've written letters to Georgia Power for families to keep their electricity on so they can maintain safe home temperatures for their infants or run the nebulizer for their asthmatic child. AVLF is also familiar with such cases. They most frequently see respiratory conditions caused or triggered by unsuitable living environments contaminated with mold, insect infestations, and other allergens. Thaler has also seen injuries from falls or faulty fixtures. "One time I had a client who leaned against a wall and her hand went right through it," Thaler remembers. The impact of living conditions extends beyond the home itself and includes the surrounding environment. Poor physical environmental conditions, such as crowding, substandard housing, elevated noise levels, and elevated exposure to noxious pollutants and allergens adversely affect health.[17] Such conditions also create constant stress, festering like a low-grade fever.

If housing and neighborhood conditions are so despicable and even toxic, why not leave? Thaler explains that the reasons are varied. One elderly woman didn't want to leave because she could walk to her doctor without crossing a street. Another mother wanted to stay in the rundown house she was renting because her daughter had a learning disability and was finally in a school with good support. Often, it is because moving is expensive. Moving requires a security deposit, startup costs for utilities, and moving fees. Finally, and perhaps most tragically, moving does not necessarily offer improved conditions given the severe lack of low-income housing across America. "A lot of people prefer the devil they know to the devil they don't," says Thaler. The cockroach infestation and rotting walls may be better than the toxic mold and broken AC down the street. Without additional affordable housing options, people are forced to stay in places that are literally making them sick. Yet, as reflected in Tre's childhood experience, even in less-than-ideal housing situations, strong social cohesion can be protecting and foster positivity. Demolishing historic neighborhoods and relocating people is not always a solution. An increase in affordable housing options, improved protection of renters, and community structures to promote safety and cohesion, such as sidewalks and parks, can make an immediate impact in our neighborhoods.

CHAPTER 8

THE CHALLENGE OF GETTING WELL

HEALTH CARE ACCESS IN THE UNITED STATES

Of all the forms of inequality,
injustice in health care is
the most shocking
and inhumane.

MARTIN LUTHER KING JR.

BREANNA

The first time I delivered the dreaded cancer diagnosis was on a hot Thursday in August. I had been a nurse practitioner for three weeks and was frantically learning everything years of schooling hadn't taught me. The small clinic looked like an abandoned building from the outside and housed two exam rooms, a waiting area, one bathroom, and a makeshift office area. I had joined the nurse practitioner who had founded the clinic, and after years without a vacation she decided that after two weeks of orientation I was ready to be on my own. I had finished seeing patients, my coworkers had left, and I sat among a stack of charts and lab results. I had gotten to the clinic early that morning as always to look through the faxes and scan each report. I already knew she had cancer, and the diagnosis loomed over me all day. She had never met me. She didn't even know my name. She was going to receive her breast cancer diagnosis from a stranger—a stranger who had three weeks of experience and no idea how to break this kind of news.

I started with a few easy calls: "Your labs look good"; "Your blood sugar is still running too high." I looked over her report one more time in case I had misread the results. I repeated the words "intraductal carcinoma" out loud in the silence of the deserted office and made the call.

The phone rang two times.

"Hello."

"Hello. Is this Ms. Emerson?"

"Yes, who's this?"

"My name is Breanna. I'm a nurse practitioner at the clinic. I work with Ms. Foster, who you've been seeing at the clinic."

"Ok . . ."

"Recently you had a biopsy done on your breast. Ms. Foster is currently out of the office. I know you don't know me, but I wanted to make sure your received your results right away."

She remained silent.

"I'm so sorry, but the biopsy found cancer cells."

She screamed. I remember her saying no over and over as words abandoned me. I remember her crying.

"How am I going to afford treatment?" she sobbed.

I was prepared for this question but the tragedy of it stung with a new intensity. She didn't ask me if there was treatment because she already knew that there was. She had already heard of cancer survivors and strides made in breast cancer survival. Her question was whether or not, in a city filled with resources, she would have access to the treatment that could save her life.

"Will I survive?" is the question of all cancer patients regardless of race, gender, or socioeconomic status. The answer, however, has everything to do with such factors. Ms. Emerson was completely rational in her fear that her lack of health insurance would negatively impact her access to treatment and ultimately her survival. People without health insurance are less likely to receive health care, more likely to be diagnosed in more advanced stages of diseases, and more likely to die than those with insurance.[1] One quarter of adults without health insurance report going without needed health care in the past year, compared to only 4 percent of adults with coverage.[2] In another survey, 41 percent of adults reported skipping doctor's appointments, forgoing recommended treatment or diagnostic studies, or leaving a prescription unfilled due to cost.[3] I see the consequences of this regularly in clinical practice. It is the man who was told three years ago at a health fair that he might have diabetes. He sits in my office with a blood sugar of 400 and a necrotic foot ulcer so extensive I am forced to send him to the hospital praying an amputation might not be necessary. He tells me he didn't come sooner because he not only had to pay for the health services but couldn't afford to lose an entire day of work. Now he will be losing much more than that.

Even once those without insurance seek care, their outcomes are not the same as those with insurance. Uninsured individuals have a higher mortality than insured individuals. Even when controlling for factors like age, unemployment, smoking, alcohol use, weight, and health status, people without insurance have a significantly higher mortality risk than the insured.[4] In a study of twenty-three million hospitalizations of children over a period of eighteen years, uninsured children were more likely to die than children with insurance.[5] Tragically, over the eighteen years of the study, 39,323 of these deaths could have been prevented by providing health insurance to all children.

People without insurance are also more likely to forgo preventive care. I will always remember the day Mac came to Good Sam a few years ago. He was a new patient and didn't have an appointment, but my last patient had not come so I agreed to see him. He was a tall, thin black man in his late forties with a pleasant disposition. He came in about his reflux, hoping for a medication to prevent the burning in his chest after each meal. He was uninsured and hadn't had any type of checkup in the past decade. His exam that day was relatively normal, with the exception of his elevated blood pressure. Reading 170/100, his blood pressure was well above the recommended 120/80. However, this is common in my practice given that many patients go for years without care and are rarely up-to-date on recommended screenings. His urine and electrocardiogram were normal, and I recommended he let me draw some labs so we could also treat his blood pressure. He agreed and left with a treatment plan and some medication for his reflux. The next day the lab called me with urgent lab results. Mac was in renal failure. With a creatinine of 41, his kidneys were almost completely nonfunctional. As a result, electrolytes like potassium were at critically high levels. I stared at the labs for a minute as I had never seen levels like this. Near-death patients in the intensive care unit have better lab values than this. I picked up the phone and dialed his number. I was connected immediately to voicemail and left an urgent message. Fifteen minutes later I had still not reached him.

He might be dead, I remember thinking. *I should drive to his house or have an ambulance go to his house.* I dialed the emergency contact number and was relieved when his brother answered the phone. His very confused brother explained to me that Mac was in a probation program during the day and did not have access to his phone.

"I know this sounds crazy," I explained, "but I need you to drive over there, pick him up, and go directly to the emergency room. There might not be much time." Fortunately, he was convinced, picked up his brother, and within three hours the hospital started dialysis. Several months later, after surgery to stent the tubes connecting his bladder and kidneys, he recovered with only minor kidney damage.

The story has a happy ending, but I've always wondered if his long hospitalization might have been prevented with an annual physical and blood pressure check. Even more disturbing is the question of what would have happened if the scheduled patient had shown up that day. His elevated potassium might have stopped his heart before the reflux would have brought him to an emergency room.

Preventive care can be lifesaving but is often unattainable for people without health insurance. Uninsured women receive mammograms to screen for breast cancer at half the rate as women who have health insurance.[6] One study of women who had not received recommended mammography screening found that over half stated that the reason was due to lack of insurance.[7] Similarly, cervical cancer and colorectal cancer screening rates are lowest among those without insurance.[8] Between 25 percent and 40 percent of disparities in the use of preventive services can be explained by differences in insurance coverage.[9] Health insurance increases the likelihood of screenings, health education, and early detection, preventing health problems and costly interventions.

Lack of health insurance is not the only barrier to accessing health care. While ensuring everyone has health insurance is an important first step in eliminating health disparities, insurance does not equate access. I have patients with insurance who cannot afford their copays. At Good Samaritan, patients without insurance pay on a sliding-scale fee according to their income. Even at reduced costs, patients may be forced to choose between food and health care. Others lack transportation to appointments or do not have anyone available to interpret for them in order to see a specialist. One mother told me that even though her mammogram would be free, she could not get to the hospital and back from her children's school, which was located far from the only hospital where the mammogram would be free. She had no one else who was able to care for her children, so the mammogram would have to wait until school was out for the summer.

Lack of health care access can also be a result of provider shortages. The United States has less primary-care providers than required to meet the health care needs of the population. The need for primary-care providers is growing more rapidly than physician supply due to population growth and aging.[10] Without intervention, the United States will experience a shortage of 20,400 full-time primary-care physicians by 2020.[11] Provider shortages are also not evenly distributed geographically across the United States. The US Department of Health and Human Services Health Resources and Services Administration (HRSA) designates health professional shortage areas throughout the United States based on population demographics and provider supply. Currently, 59.5 percent of those areas are in rural America.[12] Essential health care services are not convenient or close. For example, less than half of women in rural America live within a thirty-minute drive to the nearest hospital providing prenatal care.[13] Access to health care can be limited by location, transportation, language, provider supply, and other nonfinancial barriers.

Finally, the current US health care system is limited by our approach to health care. The US health care system is largely reactive in nature, placing resources in treatment and end-of-life care over prevention and health education. Early diagnosis of diabetes and appropriate management in a primary-care setting can delay or prevent expensive and deadly complications such has kidney failure and heart disease. Treatment of urinary tract infections and sexually transmitted infections in an outpatient setting like Good Sam prevents hospitalization for systemic infections. Compared to other economically developed countries, the United States performs well in acute care in hospital settings but not very well in preventing hospital admissions, particularly for people with chronic diseases.[14] Countries with strong primary-care systems have lower mortality rates.[15] Health care has its greatest potential for decreasing mortality when it is affordable and accessible to people prior to getting very sick.

I understand that in a resource-limited environment it seems reasonable to save someone near death rather than provide an affordable Pap test to all women. However, with population growth and increasing health care costs, the question is no longer "How can we save the sickest people?" but rather "How do we keep people from becoming so sick?"

Doreen primarily relied on the ER for care before coming to Good Sam. She was unemployed and living in a transitional housing shelter while trying to reenter the workforce. Her blood pressure was elevated and she hadn't had routine screenings such as cholesterol and diabetes checks, a Pap test, or a mammogram in at least a decade. Knowing her housing situation was transient, I decided to make the most of our time together and catch up on as many recommended screenings as possible. A Pap test is a simple screening for cervical cancer and is highly effective because cervical cancer is extremely curable if detected early. Most often the abnormal Pap tests I see show precancerous cells, which normally lead to close monitoring. In Doreen's case, her Pap was abnormal with high-grade cancer cells. One of our volunteer gynecologists performed the biopsy in office, confirming cancer. The entire diagnostic process cost us less than $100. We were able to refer her to a local hospital for treatment. An outpatient surgical procedure removed the cancerous tissue, and testing showed the cancer had not spread. Doreen now receives a Pap test every six months at Good Sam. This screening test probably saved Doreen's life or at least prevented a major abdominal surgery followed by chemotherapy. The $100 we spent in diagnosing her saved the health care system thousands of dollars. Primary care and preventive medicine save lives and control health care spending.

Excessive morbidity and mortality can also be the result of lack of access to mental-health treatment. Currently, an estimated 18 percent of adults have some type of mental-health illness from depression to substance abuse to anxiety disorders.[16] The current supply of psychiatrists is not sufficient to meet this need. Seventy-seven percent of counties in the United States do not have enough psychiatrists to meet the demand.[17] Even when psychiatrists are available, people may not be accessible due to health insurance plans without appropriate coverage for behavioral health or the growing number of private psychiatrists working on a cash-pay basis. Even if psychiatrists were plentiful, stigma and fear surrounding mental health create barriers to care. One single mom with two young children hesitantly opened up about her years of severe depression.

"Don't write it in my notes," she requested. "I don't want any treatment because I don't want anyone to take away my children." We have daily counseling appointments at Good Sam, and I could have started an antidepressant, but she had suffered silently for years due to a fear that society would deem her unfit to parent. These stories are commonplace. Mariam and Sam, who work as certified peer specialists and have lived experience of mental-health illness, describe how difficult the diagnosis process can be.[18]

"You hear things like, 'You'll never have a family' or 'You'll never work again,'" Sam explains. "It's 'You can't, you can't, you can't . . .'" Would we tell this to a diabetic patient or someone newly diagnosed with elevated blood pressure? I have other patients who have sought treatment and left discouraged. Without insurance many had stopped and started treatment without finding an option that made them feel better. For others, medication wasn't the solution, but no one discussed the many nonmedical options. When such options aren't available, patients experience disability, decreased quality of life, and poor health outcomes. Suicide is the tenth leading cause of death in the United States and the second leading cause of death in individuals between the ages of ten and thirty-four.[19]

With the challenges of access, affordability, and mental-health parity in treatment, clinical care sometimes feels like treading water. I can help my patients stay afloat, but how do we get rid of what's drowning them in the first place?

CHAPTER 9

THE UNMAKING

ON MOVING OUT AND REBUILDING

This is where the walls gave way
This is demolition day
All the debris and all this dust
What is left of what once was
Sorting through what goes and what should stay

NICHOLE NORDEMAN, "THE UNMAKING"

VERONICA

When the day came in April 2016 that we finally moved out of our home in Southwest Atlanta, nine years after moving in, all the kids in the neighborhood we had known and loved were gone. Their families had worked hard to get them to a better place, and they moved out as soon as they could. While we missed them, we were happy for them too. For all the times our house was packed with children and teens helping us do whatever we were doing, on this day it was just us and my parents, a whole mess of moving boxes, and some homemade chicken salad dropped off by Rose, a neighbor and fellow community volunteer.

When we first started exploring the possibility of moving into a low-income neighborhood, someone told me the appropriate motivation for doing such should be that "you just like living with poor people." I know what he meant, that we should mimic Jesus' comfort level and inclination to hang out with those society casts aside. But the reality is, I always found myself not "liking it" because I was either too burdened in my heart for their circumstances or suffering myself from the effects of an impoverished environment. But I wasn't alone in this sentiment. I observed that people living there didn't like it either and did their best to pursue upward mobility. When I first told Rose, an older African American woman who owned and lived in her home for decades, that we were moving her response was "please take me with you." Shortly after we moved out and the housing market began to recover, she sold her home and left too. It's possible to romanticize people living in poor neighborhoods and cite a comfort level with "less" that is simply not true. They want and need the same things everyone desires and are equally frustrated by the crime, drugs, gangs, and disrepair of their homes. Needless to say, when moving day came, I was beyond ready to go.

But with all this, the excitement of finally moving to a nicer neighborhood was tempered by a bittersweet nostalgia and deep spiritual guilt. We had lived in this neighborhood through most of our marriage and the birth of our first child. We had poured sweat and tears into renovating our old home and building relationships with the residents. Tired and confused by the outcome of the experience, we left with a sinking feeling that perhaps we were the ones Jesus talks about when he condemns those who "put a hand to the plow and look back" (Luke 9:62). A lot of our identity had been wrapped up in being urban missionaries, and I had hung my spiritual hat on the fact that I had done something brave for God. When we moved out, we understood all the reasons people choose comfort and security. It was sobering, humbling, and disorienting.

When we first decided to move into the neighborhood, the Christian Community Development Association was saying that if you move to the inner city, you should commit to going for at least five years. The idea was that leaving before that time would prevent you from really making a change and instead send a negative message to the community, doing more harm than good. A couple years in to the experience, they changed it to say fifteen years was a better minimum. Even though Eric and I had initially committed to five years and had no solid expectations for outcomes, when we left after nine years it felt like it had not been enough. I don't know what "enough" would have looked like; I just know that deep down I felt like for the sacrifice and effort involved, community development work should be more effective. Vacant homes should be repaired and lived in, streets should be cleaned up, crime should be down, and policies should change. I was disappointed that for all the time, financial investment, and advocacy, our neighborhood still looked exactly the same. We were seeing the limitations of community development work and that this alone could not fix the pervasive policy, education, and economic issues at hand.

So as we packed up the last box and loaded it onto the U-Haul, I looked back at our home. Both good and hard memories came flooding back, and my exhausted heart sighed with relief. Eric and I were leaving as different people than when we first arrived. I left with severe anxiety, major depression, and recurrent panic attack episodes. Eric left with panic attacks too, along with high blood pressure and heart palpitations. We both left with psoriasis. Yet, even though I knew we were doing the right thing for the health of our family, I was grieving the loss of a vision and hope that community development alone could repair communities in a holistic, lasting, and scalable manner. As we pulled onto the highway, I turned around to look at the exit I had taken thousands of times to get home and thought, *There has to be a better way to restore our communities.*

BREANNA

I remember standing in the shower, the water just shy of burning my skin as if the water could wash away the pain of the day along with the germs. The day had started before the sun came up, and it was dark when I got home. A heavy stack of charts needing completion accompanied me home and sat waiting on the couch. My ever-supportive husband, who had moved across the country with me and supported me through two master's degrees, had waited on me for dinner. I was a year and a half into my career. We had bought a house, and he was thinking about the future. His question, while asked with honesty and kindness, seemed like a needle poking a balloon ready to burst.

"Can you do this forever? How will we do this with a family?"

I had always been a girl with a plan, and I was living the plan. Yet I found myself angrily shouting at God in the shower. This wasn't working. I was exhausted and questioning how a job I loved so much seemed incompatible with every other aspect of my life. But what made it even more devastating was the question I couldn't seem to let myself ask: *Was all of this making a difference? Is anyone actually healthier? Am I spending every waking hour sticking Band-Aids on stab wounds and treating epidemics with aspirin?*

I recall this moment as a realization of my personal limitations, but more significantly a realization of the limits of my profession. I had spent years learning how to prescribe the right cure for the right illness only to learn I couldn't prescribe the things my patients needed the most. I couldn't write a prescription for a new home for the young teen who was being repeatedly raped by her dad's friend who had moved into their apartment. The pharmacies couldn't dispense a job for the schizophrenic mother who wanted desperately to maintain custody of her beloved little girl. I couldn't create a family for the recently immigrated young man as I told him his only chance of survival was a kidney transplant due to an undiagnosed genetic flaw.

I had spoken, written, and advocated on the importance of health insurance for everyone only to learn health insurance wasn't the greatest need in my community. Even low-cost health insurance isn't helpful if you can't make rent. Four-dollar antibiotics cost a lot more if you miss a day of work to get a prescription for them. Medical care simply isn't enough to make people healthy. As the months went on, this reality became ever more apparent, and my questions grew more difficult to shrug off. *Am I helping people live longer or helping them die a little slower? Is there even a difference?*

I went to bed devastated that night. The charts were finished and waiting to be hauled back to the clinic before dawn the next morning. I'd unlock the clinic door in the morning and patients would fill the waiting area. I would ask the right questions, determine a diagnosis, and develop an affordable treatment plan. Yet patients wouldn't leave healthy. They would return to the communities, jobs, social constructs, and political systems that were making them sick. And when I locked the door, set the labs out for pickup, and piled charts in the passenger seat of my car, I'd drive the block back to my house thinking: *There has to be a better way to make people healthy.*

PART TWO

HOW WE GET WELL

CHAPTER 10

A NEW BEGINNING

WHAT WILL MAKE OUR NEIGHBORS HEALTHY?

When you are in the middle of a story it isn't a story at all, but only a confusion; a dark roaring, a blindness, a wreckage of shattered glass and splintered wood. . . . It's only afterwards that it becomes anything like a story at all. When you are telling it, to yourself or to someone else.

MARGARET ATWOOD, *ALIAS GRACE*

As winter comes to an end, the urban farm at the Good Samaritan Health Center looks like a field of mulch and remnants of last year's harvest. The farmers spend long hours manipulating Georgia clay into fertile soil and planting seedlings. Yet, other than their dirt-covered boots and callused hands, there is not much to see. From a glance the farm looks as bare as it has all winter, and yet tiny plants are thriving, ready to produce thousands of pounds of produce in a few short months. Our neighborhood can feel much the same way. A local food business opens to provide healthy options in the community. The number of homeless Atlanta residents decreases for the first time in years. Patients at the clinic are cured of hepatitis C through new prescription-assistance programs. Signs of health, like the little seedlings on the farm, offer a vision of what can be, and like the natural regeneration of the earth in springtime, health is what we were made and designed for. Health equity can be a reality.

Yet, as our farmers will attest, creating a thriving farm requires months of labor. Making our neighborhoods healthy requires the complex yet critical work of dismantling systems of oppression and creating the social structures needed for health equity. Across the nation this work is being done by groups of community members, local leaders, advocates, professionals, and volunteers every day. While poverty, racism, unemployment, and lack of housing and health care have devastating health implications, economic opportunity, reconciliation, employment, and affordable, quality housing and health care can transform communities and reduce mortality. There is hope. Part two contains stories, exemplars, and strategies for healing, beginning with our own.

VERONICA: THE PROCESS OF HEALING

By the time we moved out of Southwest Atlanta we were running on fumes. Our financial situation was a constant source of tension since our house was still underwater as a result of the 2007 financial crisis and the poor economic prospects of our neighborhood. We are the type of people who like to do things right, with proper planning, and diligent execution. We consider ourselves responsible and kind. Yet the more we tried to "do the right thing" the more we realized how broken everything had become and how much bad news followed us everywhere we went. It felt like we had chosen downward mobility in life. Compared to our happy first year of marriage, which we spent on the nice side of town, we were now angry and sad and finding it harder to make life work. What was happening is the reality that *place matters*—neighborhood environmental factors, from local economic opportunities to social interactions with neighbors, to the physical environment, to the local grocery, all affect the individual health.[1] This time my family was feeling the effects.

After leaving Southwest Atlanta, things got much worse before they got better. I'm not an expert on posttraumatic stress disorder, but I'm pretty sure we dealt with that for our first six months in the new home. We were in a better environment, but we were still sick. It took a solid year before we gained control of our lives again. This was a confusing and bleak time, and some days the only thing that got me through was believing, despite the present darkness, that light would eventually break through and there were better days ahead.

In one particular visit with my psychiatrist, I opened up about the experience of living in the neighborhood. The doctor nodded and explained that this type of prolonged stress experience can indeed cause trauma to the brain and lead to depression. The doctor was not nearly as surprised as I was that my brain was injured and taking a long time to bounce back. Thanks to the compassionate health care I received, family support, and new neighborhood, life started to improve. I was receiving appropriate mental-health treatment in a safe environment and my brain slowly began to heal.

Once we were settled into our new home and new community, I immediately noticed the differences in our quality of life. From the start my home, about eighty years newer than my previous house, had been well-maintained by the previous owners and had very few repair needs. It was obvious that the family who lived here previously loved the place and made it home. The property across the street was a beautiful community garden that was active every day with young families gardening, walking, and playing together. We weren't afraid of the outdoors anymore, nor did we fear daily for the safety of our child. We moved in late April 2016 and by the Fourth of July we had already been invited to a neighborhood party on our street. There were so many happy young families with small children and dogs, just like us, in similar stages of life. We immediately felt like part of a close-knit community. In fact, the day we viewed the home before deciding to purchase it, we met three friendly sets of neighbors who gushed about how much they loved the neighborhood. It was such a stark contrast from Southwest Atlanta that it almost felt like an out of body experience.

Within the first few weeks I also noticed structural differences about our new neighborhood. They were far fewer vacant homes, and the ones that were vacant were receiving quick attention from the city code-enforcement office whenever a complaint was issued. Unfortunately, because we were still living in the city, crime was occasionally an issue, and within the first three months in our new home there was a shooting one street over and occasional car break-ins. Even though we were still experiencing crime issues, there was a significant difference in the way the neighbors and police station responded. After the shooting, an emergency safety meeting was called by the neighborhood association, and so many residents showed up that we could not fit in the church conference room where the meetings were normally held. A host of city officials and leaders from the police force attended the meeting and expressed sincere concern and immediate action steps to better patrol the area and prevent future issues. The response was unlike anything I had experienced in Southwest Atlanta, where most incidences of crime went without any such response from the community or the police. It was a strange feeling to be sitting in the middle of a tense meeting on neighborhood safety and feel strangely comforted based on the fact that people cared and city services were engaged.

There are other things, like the difference it makes when you live in a neighborhood where people have the will and resources to maintain their homes, plant flowers in their front yard, and walk their dogs, and young mothers can push their babies around in strollers at dusk without feeling concerned. I hear a lot more birds chirping, kids laughing, and neighbors celebrating. Cars rarely speed down the street, and when they do you can count on a mom hollering after them or calling them out later that day on the neighborhood Listserv. The smell of gardenias, rosemary, and roses is always in the air. No one needs bars on their windows, and our dogs are less on edge all the time—they've actually grown bored and need interactive dog toys to pass the time when we are at work. We spend many evenings eating dinner on our big front porch, talking with passersby and watching birds of all types eat from the feeder and nest in our trees.

A neighbor quickly added me to the neighborhood moms' text thread, and now I have all these new friends with kids my daughter's age. People watch out for each other, and even though they don't always agree; you can plainly see they are trying their best to make the neighborhood a great place to live. Sitting on the front porch one evening my father-in-law referred to it as "the garden of Eden," and that's how it feels. There is a sense of shalom and peacefulness in daily existence that we have not experienced in over a decade. We are like flowers that have been transplanted from a neglected garden to one of a master. With better sun, soil, and nutrients we are coming back to life, but it is a slow and messy process. It took over a year, but I finally feel safe falling asleep at night and have shed the ever-present fear that I lived with for so long.

There came a point when it was hard for me to write this book. Perhaps not for the reason one might expect. It was not because the memories were too painful, but rather the opposite—once my family was living in a healthier environment, I started to forget. One year in, my husband and I were healing emotionally, our marriage was improving, our finances were starting to recover, and the weight of feeling scared at home was lifted. My complexion, which broke out badly near the end of our Southwest Atlanta experience, was clear and healthy. I was making art out of joy rather than pain, and drinking to celebrate rather than to numb. The change in neighborhood environment made a drastic difference in all aspects of my health and behavior.

So there was a lull in my writing until one night I was lying awake in bed wondering how I was going to finish the story. I closed my eyes and traveled back in my mind to remember a typical walk in the neighborhood. Before I knew it my breath had grown short, I was frowning, and my right hand covered my heart. I was imagining being back on my street, and the memory alone caused panic and anxiety. I remembered there were still many people I loved living in that environment and felt a deep sense of urgency to finish what I had started.

I didn't want the answer for my family to be moving out. Community development training taught me that moving out too soon can cause more damage than never going at all. We wanted to make a difference, and we certainly didn't want to be those people who gave up. In fact, we stayed years too long trying to prevent that narrative from unfolding. This is why it was crushing when we finally left—we felt like such failures. In reality though, what we were doing (unbeknownst to us at the time) was proving the power of social determinants on health status and the heavy lift it requires to heal broken, languishing people.

This all started to make sense to me when I took a fundraising job at the Good Samaritan Health Center in February of 2015. At the most basic level, I was now in a culture of watching talented clinical providers diagnose sickness and develop treatment plans to heal their patients. The process that the medical community uses to identify disease and implement solutions to reverse illness started to make a lot of sense—even when applied to larger issues that I previously considered exclusively moral or ethical problems. For example, the Avielle Foundation, a nonprofit established in honor of six-year-old Avielle Rose Richman, who was murdered on December 14, 2012, in the Sandy Hook Elementary School massacre in Newtown, Connecticut, is committed to brain research and figuring out why people commit acts of violence. Avielle's father, who is also the organization's CEO, is a neuroscientist by trade and believes that even violence can be studied and treated as a disease. Further, he believes the more we know about the brain's chemistry and structure, the better we can prevent and intervene early enough to heal individuals whose brains have endured chronic risk factors and are on a path toward violence and aggression.[2] I think this perspective in the scientific community—viewing traditional "sin issues" with less stigma and more as a disease—is an important paradigm shift. If we see sin issues as primarily a personal failing or spiritual rebellion, our compassion and solutions may be limited. However, if we take the posture of a clinician and seek to understand the problem, identify what's going on, and develop personalized treatment solutions, we will offer dignity to the patient and achieve better outcomes.

It takes so much to get one person to a place of health and wellness, even with abundant time and resources. The idea of lifting the health of an entire community, sustained over time, on minimal resources is nothing short of a miracle. It's a "feeding thousands with a few fish" kind of task. Where would an activist or an organization even dare to begin? There are groups that have started with housing, education, safety, or jobs. Which one is most effective? Jim Rouse, the founder of Enterprise Community Partners, a national organization committed to improving communities and people's lives by making well-designed homes affordable, says "We can't just settle for doing some housing, finding some jobs or building some human support systems. We must do it all—decent housing in decent neighborhoods for everyone."[3] In other words, the Christian Community Development mantra is true, "If you're going to care *at all*, you must care about *it all*."

Hence it must be that those individuals and organizations engaged in place-based work must function as a web (or safety net) of resources instead of focusing only on their particular service niche. To avoid mission creep, organizations can still prioritize doing what they do best while asking the question, What can we do that will have the greatest impact on the health of this community? It is a strategic question for every group to ask since the health status of the community will ultimately dictate how effective their organization's interventions will be. If all service organizations viewed health beyond the absence of disease and embraced a concept of health equity as the collective goal, we would all be more successful in accomplishing our individual goals. Health is the goal we all want and share—whether it be financial health or mental health, language nutrition or food nutrition, we want the poor to feel better and live better.

It's not as big of a paradigm shift as we might think. For example, it's a nonprofit serving women being rescued from sex trafficking establishing a referral process with a clinic to make sure the women get appropriate ob-gyn care and counseling. Even better, it's this same organization training the local clinic or hospital doctors to recognize signs of trafficking in victims so they can proactively offer support. It's a legal aid society, upon realizing that many of their pro-bono clients deal with respiratory issues resulting from poor housing conditions, connecting with health care providers to ensure there is documented evidence of environment-related asthma as well as prescribe treatment. It's a job training volunteer celebrating a young man's GED attainment with a high five, then pausing to ask the extra question, "How is your health?"

The reverse is also necessary. Large health care systems must have knowledge of and direct links to social services. Kaiser Permanente has instituted a "life situation form" that patients complete when they visit their primary doctor. It includes questions like, Do you have trouble paying your bills? Do you feel safe at home? Do you have enough to eat? The Kaiser people have realized social determinants are inextricably linked to health outcomes. Kaiser goes further than just collecting this information. Not only do they assign a patient navigator who assists with guiding patient care within the Kaiser network, but they also aid in connecting the patient to social services. After Kaiser started offering patients this sort of support, one study found a 40 percent reduction in emergency room use.[4] A reduction in ER use is good for all of us. It saves health care dollars and reduces the strain on our local hospitals.

In action, this is what the nonprofit world has started to refer to as "collective impact," which is a framework to tackle deeply entrenched and complex social problems. It is an innovative and structured approach to making collaboration work across government, business, philanthropy, nonprofit organizations, and citizens to achieve significant and lasting social change. It is based on the belief that no single organization or program can solve the increasingly complex social problems we face as a society. The approach calls for multiple organizations or entities from different sectors to abandon their own agenda in favor of a common agenda, shared measurement, and alignment of effort.[5]

An example of this is when the local leadership in our new neighborhood rallied residents, city council members, business leaders, and the Park Pride nonprofit organization to engage in a "Park Visioning Process" to improve our local greenspace. Park Pride offered their organizational resources to provide landscape architects, run surveys, develop land-use plans, and solicit neighborhood feedback through a transparent communication process. After multiple neighborhood town hall meetings, a final park design was voted on and Park Pride provided a professional packet that the neighborhood could use to solicit funding to turn their dream into a reality. A volunteer committee of community residents signed a participation contract that was agreed on at the beginning of the project. Updates were sent regularly to all community stakeholders via email, signage, and door-to-door conversations.

Without the Park Pride team and their specific knowledge about park design, our neighborhood residents (despite their best intentions) would have no idea where to begin on creating a new park. Park Pride's expertise, combined with the passion and input of the community and matched by the economic interests of investors and city council made the difference. A large grant was awarded to fund phase one of the park renovation. This is a very small-scale example of collective impact, but it shows the power that is possible when different groups with different agendas come together and all agree on a problem and a solution for change. As John Kania and Mark Kramer, the founders of the collective-impact approach explain, "We believe that there is no other way society will achieve large-scale progress against the urgent and complex problems of our time unless a collective impact approach becomes the accepted way of doing business."

BREANNA: A THEOLOGY OF HOPE

In my first year of nursing school I was invited to attend a panel discussion on HIV policy and funding. I drove forty-five minutes from my small college town to the large university in the city and sat among people with years of public health experience. One young, charismatic man shared his story of living with HIV and the ways that having access to medication dictated his every decision. He spoke about how he loved his home but would move if he couldn't access his medication. The event was not a religious one but I remember him pausing a moment and then saying, "I just don't understand how the people who want me to come to their churches are the same ones who are doing everything they can to cut the funding that is keeping me alive." I don't know what experience this man had with church or Christians, but he had observed that people who were outspoken about their faith were often the same people who publicly opposed policies most critical to his health. During the emergence and expansion of HIV treatment, there certainly were Christian advocates and congregations supporting funding and treatment for individuals living with HIV. However, this man had observed a correlation between people's advocacy against policies enabling him to afford treatment and their religion.

The audience was silent with little reaction except for numerous nods. The panel went on, but not for me. I cried the entire forty-five-minute drive home. How was it that this man saw Christians as the people who were sentencing him to death? How could it be that I felt such excitement in my future identity as a health care provider and such shame in my Christian identity at that moment?

I remember this moment as the start of an ongoing struggle. Despite my love for God and dedication to staying in a church community, there was a huge disconnect between my professional passion and my religious background. I knew many wonderful Christian people but longed for conversations about the intersection between God's love and social justice. I spent the week surrounded by the sick and hurting and recognized more and more how society contributed to and ignored their suffering. But on Sunday, no one talked about that. I kept asking myself, *Does the church care that our neighbors are dying?*

Let me pause here to say that many congregations do care about the lives of their neighbors. Over time I have found faith communities of diverse beliefs who seek social justice and care deeply about the suffering and injustice of this world. Many such faith communities have been seeking reconciliation, healing, and justice for decades in ways far beyond my own encounters with social justice and the gospel. The privilege afforded me by my social status (upper-middle class), race (white), and economic resources (education, personal car) allowed me to reach college before seeing the disconnect between my faith and lifestyle.

JESUS AND SOCIAL DETERMINANTS OF HEALTH

Coming to work for the Good Samaritan Health Center was a source of healing for me. I started experiencing a connection between my faith and daily work. I found myself surrounded by people who saw their vocation as a response to God's love in their lives. Our staff, volunteers, and donors come from various racial, ethnic, and socioeconomic backgrounds. They attend a wide range of congregations and differ in beliefs from theological issues to political parties. However, we all agree that God wants his creation to be healthy. The

Gospels show Jesus healing, serving, and being with people, often the ones most ostracized by society. Consider the following examples.

On more than one occasion Jesus heals those suffering from leprosy (Matthew 8:1-4; Luke 17:11-14). Leprosy, now understood as a curable infection caused by *Mycobacterium leprae*, was at that time seen as a horrible punishment from God. The disease is debilitating as the bacterium slowly attacks the nerves causing loss of sensation, skin ulcers, and the disintegration of fingers and toes. In biblical times, lepers were forced to live outside the city and were never touched. Yet Jesus touched them and restored their broken skin and failing bodies.

Once when Jesus was traveling, a woman who had been bleeding for twelve years touched the hem of his garment in hope of healing (Luke 8:43-48). She was likely in great physical distress. Without the interventions of modern medicine, she would have been fatigued and weakened from anemia. However, her psychological distress was likely even more devastating. In biblical times, bleeding rendered a woman "unclean." Being unclean would have meant twelve years without physical touch, and no man could be with her, leaving her on the lowest rung of society.

When Jesus asked who had touched him, the crowd withdrew in fear and the woman fell trembling at his feet. Jesus met her with grace saying, "Daughter, your faith has healed you. Go in peace." I have had several patients suffer through years of heavy bleeding due to fibroids. One saved for three years to afford a hysterectomy since she did not have insurance. Another patient is still waiting and requires blood transfusions in the ER every couple of months. She is in the United States illegally and has had difficulty finding assistance due to her legal status. For Jesus, healing wasn't limited to those who were rich or socially clean. Yet wealth, insurance status, and legal status determine the health care options for many of my patients.

Another time, Jesus and his disciples came upon a man who was born blind (John 9:1-7). His disciples asked him whether it was the man's sin or his parents' that produced his blindness. Jesus replies, "Neither this man nor his parents sinned." In other words, physical aliment is not a punishment for sin. The abundance of suffering in this world is neither fair nor is it God's desire. I think of many of our dental patients who are missing half of their teeth and the half remaining are rotted. We can ask what caused it: Drug use? Lack of proper nutrition? Domestic abuse? Time in prison? A lifetime without dental care? But the explanation isn't what matters. They are God's children, equally deserving of a functional mouth and beautiful smile.

Jesus didn't have to heal anyone. He was clearly a gifted teacher and preacher. He could have performed only miracles that benefited large crowds and impressed the skeptics. However, over and over again we see Jesus healing people. These stories illustrate that the present world matters to God. What has always intrigued me about Jesus' healings is that they were not limited to physical health. Jesus consistently disrupted social structures, changing not only people's physical condition but their social status. Jesus, thousands of years ago, cared about the entire person and healed people in a way that today we would consider addressed the social determinants of health:

When Jesus healed the lepers, he changed their position in society. Cleansed from their physical disease, these men and women could live within community, work, marry, and have children. Their numb extremities and skin lesions were external manifestations of a life marred by isolation and discrimination. Today, a combination of strong antibiotics can cure leprosy. Reversing isolation, discrimination, and shame requires nonclinical interventions.

I imagine the crowd surrounding the bleeding woman was expecting to see her reprimanded. Her gender alone placed her in the lowest levels of society, and her bleeding rendered her unclean. How dare she touch Jesus? Yet Jesus calls her "daughter" and affirms her as a woman of great faith. Jesus likely confused at least a handful of people in the crowd that day. Social norms would not have allowed

Jesus to heal this suffering woman. Yet he not only healed her but revealed to the crowd that those he called faithful could be of any gender and of any societal level.

The blind man, like the lepers and the suffering woman, likely found a new place in society, no longer left to beg for survival. Yet his story is even more complex in that Jesus healed him on the Sabbath, a day when no work is allowed. The synagogue leaders interrogated the healed man and his family, more concerned with keeping social rules in place than with the fact that he was miraculously given his life back. Stories like this remind us that Jesus was more concerned with the lives of his neighbors than the religious customs and structures of the day. While a strict observation of Sabbath preventing the work of health care providers would be hard to find in this day and age, some churches today both explicitly and implicitly uphold systems of oppression with health implications.

TOWARD HEALTH EQUITY

Part two focuses on the hope, modeled by Jesus' actions, that a better world is possible. Life-expectancy gaps can narrow, communities can be strengthened, and individuals can move from sickness to health when we structure society around the belief that all neighborhoods deserve to be healthy. We have written part two jointly, providing examples of innovative solutions to poverty, unemployment, food deserts, and lack of health care. These examples include local interventions such as the Good Samaritan Health Center and Georgia Works, which have a significant impact on the well-being of specific communities. We also discuss the role of advocacy through groups like the Atlanta Volunteer Lawyers Foundation and the role of local government through the example of the Office of Equity and Social Justice in King County, Washington. Finally, we explore the ability of small pilot programs like Housing First to become national movements for change. While our examples feature individuals and organizations diverse in their location, reach, and mission, at the core of each is the belief that health equity is possible.

We begin with our personal experiences at Good Sam, highlighting the way the clinic considers social determinants and expands beyond the boundaries of traditional medical care. We share the stories of patients experiencing the negative impact of social determinants of health throughout the social gradient, from those experiencing homelessness to low income renters, from the unemployed to those with full time jobs. We discuss the interventions we are a part of that aim to uniquely address the health inequities facing our patients.

We also conducted interviews with people in our local community and across the United States who are implementing creative and effective solutions to address specific social determinants. Our personal expertise lies in nonprofit development, public health, and health care. Therefore, many of the programs and interventions we feature interact directly with the health care profession. While many of the featured solutions target specific populations, such as homeless individuals, each is a model for broader interventions illustrating the way in which focused approaches addressing specific social determinants can contribute to healthier neighborhoods. For those of us working in health care, we can be leaders in confronting social determinants by leaving our medical silo to address the root causes of poor health. For those working outside of the health care system, remember that health outcomes are driven by physical environment, employment, social status, education, and socioeconomic status. Improving health outcomes is not solely the responsibility of the health care community.

We recognize that the task of addressing social determinants of health can be intimidating and overwhelming. Throughout part two we offer practical ways individuals can contribute to creating healthier communities. The commitment of individuals to make meaningful change, however small in reach, amounts to significant change at an aggregate level. For all the stories of pain, disappointment, and sickness, we have also seen restoration, healing, and progress in our community. As a nation, we have a difficult task ahead of narrowing the life-expectancy gap and reaching health equity. However, changing our understanding of health and the way it is achieved is a good place to start.

CHAPTER 11

THE GOOD SAM STORY

We don't know how to dumb it down. We don't know how to provide "B-level" service. So our patients, who may be financially needy, still get the same level of excellent care that anyone would get going to a traditional suburban private practice.

DR. BILL WARREN, FOUNDER OF GOOD SAM

The Good Samaritan Health Center is a charitable clinic in Atlanta and a powerful picture of diverse people, including staff, volunteers, and donors, finding common ground in the belief that God wants us to love people and offer avenues for healing. Through our work at Good Sam, we have experienced ways in which local organizations can address social determinants of health and how we as individuals can impact our community. The Good Sam story highlights our ongoing effort to improve health equity in our community and the ways we have gone beyond traditional health care services to accomplish this goal. In our work at Good Sam we regularly ask the question, What will make our neighbors healthy? If this question guided all of our national and local policy decisions, organizational programs, community planning, and personal choices across the United States, life-expectancy gaps would narrow and health equity could be obtained.

Through our work at Good Sam, we have met many people who care about reducing health disparities and life-expectancy gaps, including Good Sam's founder, Dr. Bill Warren. Born to an affluent Atlanta family, Dr. Warren could have chosen to do anything with his life. Yet he knew from an early age that he wanted to help heal people, and thus he embarked on the journey to becoming a pediatrician. He started his career in a busy suburban pediatric practice, and while he loved his work, he felt God was calling him elsewhere. He spent more time thinking about the disparity in health care within the city and felt strongly that the same quality of care he provided to families in wealthy neighborhoods should be available to the uninsured, undocumented, and homeless of Atlanta. Everyone thought he was crazy when he left his extremely successful private practice to operate out of a small, run-down spare room in a downtown Baptist church. The journey was expensive and hard, and at times Dr. Warren questioned the call—but he kept serving and the ministry grew. In 1998, he formed the Good Samaritan Health Center, which opened for service in 1999 in downtown Atlanta with a staff of eight and a handful of volunteers.

Good Sam was founded with the mission of "spreading Christ's love through quality healthcare to those in need." Good Sam serves a diverse population. Some patients are immigrants who have traveled across the globe and others have risked illegal entry in hopes of a better life. Many have lived in a single urban neighborhood their entire lives and have watched the city change. Some are homeless and others homeowners. Some are unemployed while others work three jobs to make ends meet. Regardless, at Good Sam the entire family can receive quality health care in an atmosphere of dignity and respect regardless of race, ethnicity, sexual orientation, or religion.

The clinic specifically serves those living at 200 percent or less of the federal poverty level, including the uninsured, underinsured, working poor, homeless, elderly, immigrants, and undocumented. Patients receive high-quality primary care, specialty care, dental care, mental-health services, spiritual care, nutritional counseling, health-education resources through the on-site urban farm and farmer's market, cooking classes, and fitness center, and access to medications at the on-site dispensary. Dr. Warren likes to refer to it as a "one-stop shop" for the whole family. Good Sam has grown over the last twenty years to a full-time staff of forty-five employees, countless volunteers, and nearly thirty-nine thousand patient encounters annually.

Since the beginning Dr. Warren, the staff, and the board of directors recognized the need to expand beyond traditional medical services in order to meet the needs of vulnerable communities. As the clinic grows, we continue to build on the goal of a social determinants-focused approach to health care delivery. This effort led to the development of the "full circle of health" model currently implemented at

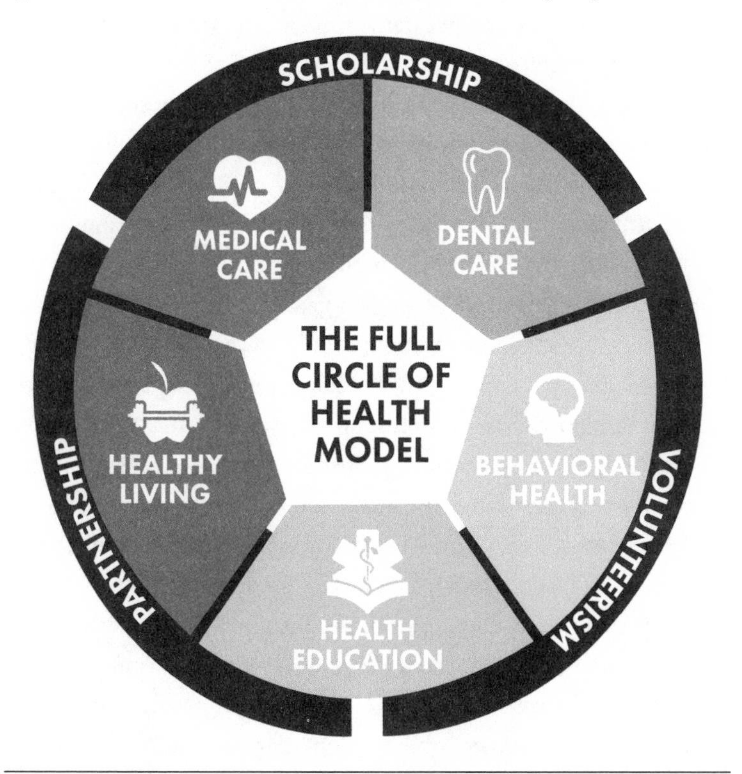

Figure 2. The Good Samaritan Full Circle of Health

Good Sam. The full circle of health (see figure 2) includes medical care, dental care, behavioral health services, health education, and healthy living tools.[1]

Good Sam's model is designed to surround patients with support and promote health rather than simply diagnose illness. If God cares about the entire person, so do we. Often a sympathetic, nonjudgmental listening ear is as therapeutic as medication. Patients can receive blood work, but they can also ask for prayer, pick up fresh produce, attend a chef-led cooking demonstration, or take a yoga class. The model works best when patients take advantage of some or all of the circle components. We make a bigger impact on health outcomes when we think of health care extending beyond the exam room.

The medical department offers primary care, including pediatrics, prenatal care, health checks, and chronic-disease management. Volunteers provide specialty care such as cardiology, orthopedics, and simple surgical procedures. The dental department provides restorative and preventive dental care. Patients might receive routine hygiene care or extractions followed by dentures. Good Sam has included behavioral health since the beginning, recognizing the importance of mental well-being on health status. Good Sam has a full-time counseling department and psychiatric services. The health-education component of the full circle model offers programs to patients and the local community to support healthy decision-making. Education opportunities include prenatal classes, diabetes group visits, kidney health classes, and a variety of lifestyle programs. Healthy living expands beyond education. Directly behind the clinic sits Good Sam's urban farm. The one-acre farm produces over thirteen thousand pounds of produce each year, which is used in the cooking classes and sold at Good Sam's daily farmer's market. Patients can learn about the nutritional value of specific foods, watch a cooking demonstration on how to prepare it, and then purchase the produce on their way out. Good Sam's on-site fitness center, run by the YMCA of Metro Atlanta, offers daily group exercise classes and personalized wellness coaching.

The full circle of health model is supported by volunteers, partnerships, and scholarship. Volunteers help staff the farm, farmer's market, prayer room, and clinic. Over forty clinical volunteers provide $250,000 in clinical services annually and allow patients to receive specialty care that might otherwise be unaffordable. Addressing social determinants of health requires support systems beyond the scope and capacity of Good Sam. To bridge this gap, Good Sam has built partnerships with organizations that provide the services we do not, such as housing assistance or job training. Scholarship refers to our dedication to clinical excellence. We strive to provide the highest quality, evidence-based care, which is usually unattainable for people without health insurance or financial resources. Our aim is to operate like a private practice and provide the level of care that those with financial resources and insurance would expect from their health care provider. As Dr. Warren likes to say, "I was trained to provide A-level care, and I don't know how to dumb it down. So that is what we're going to do."[2] True to his word, Dr. Warren is adamant about quality, from the clinicians we hire to all the chairs matching in the lobby; every detail matters. It matters because when a patient walks in we want them to see that their health is important and they deserve the very best.

FUNDING HEALTH

Because of Good Sam's high-quality care, innovative model, and holistic services, every time we give someone a tour they inevitably ask, "How do you fund this?" Good Sam is not a federally qualified health center, meaning we receive no direct government support beyond reimbursement from patients with Medicare and Medicaid. Rather, we operate like a typical nonprofit that is philanthropically funded and relies on donations to sustain and grow our operations. Our annual budget of $3.6 million is 80 percent supported by individual, church, foundation, and corporate donations, and the remaining 20 percent comes as earned income from patient fees based on a sliding fee scale and minimal insurance reimbursements. Good Sam is fortunate to have a strong base of faithful, generous supporters and a board-designated endowment that covers a portion of our annual operations.

Yet even Good Sam struggles some years to make ends meet, and as recently as 2017 it experienced a financial loss because expenses related to a growing patient base and increased quality of care outpaced our revenue. This is not an uncommon situation in the charitable and free clinic sector. Many safety-net clinics struggle to finance their operations or have a lengthy waiting list because the local demand far outweighs their capacity. Yet this should not be the case. There is strong economic evidence to support investment in charitable clinics according to a 2015 study conducted by the Economic Evaluation Research Group at the University of Georgia College of Public Health. As just one of many examples cited in the report, for every one hundred patients that visit a clinic annually, $50,000 is saved in avoided visits to emergency rooms. At Good Sam alone, which serves on average seven thousand unique individuals per year, this would be a savings of over 3.5 million dollars to the local economy, and reduce emergency room waits significantly.[3] Clinics are also extremely efficient at maximizing their resources. A 2015 study by the National Association of Free and Charitable Clinics concluded that for every dollar donated to a free or charitable clinic, the clinic can provide five dollars worth of patient services.[4] Investing in a local clinic not only makes philanthropic sense but economic sense as well. Everyone wins when the free and charitable clinic safety net is strong.

However, four current giving trends prevent adequate flow of resources to nonprofits that need it most. Money is nonprofits' oxygen. Simply put, if they don't have it they can't function. As one clinic director likes to say, "Running a clinic is like running a business. If the business goes out of business, then there is no mission." Clinics are often operating on a shoestring budget and achieving incredible outcomes despite limited resources. Imagine the impact they could make with increased support! But before this is possible, we have to overcome some public misconceptions about how to evaluate and how to best help nonprofit organizations.

In our experience in the nonprofit fundraising industry, we have watched a steady trend toward increased scrutiny on the part of the donor before making a contribution. What used to be largely handshake transactions built on trust has evolved into more formal relationships requiring increased reporting and accountability. This is, on the whole, a positive transition toward necessary transparency. However, nonprofits and donors are not always on the same page about what metrics should be used in this assessment process. Two of the most common public misconceptions regarding nonprofits are the overhead myth and the burden of evaluation.

The *overhead myth* is a term coined by social activist and fundraiser Dan Pallotta that challenges the old notion that an effective nonprofit should run on very little overhead dollars. In fact, many watchdog agencies encourage donors to look at this measure, sometimes above all other considerations. Pallotta questions this approach and offers a new paradigm: rather than reward nonprofits for how little they spend, reward them for what they accomplish. It will take major financial investments in quality staff, technology, fundraising, and sometimes risk-taking to solve the most pressing issues facing our world today.

We heard Dan Pallotta speak at a conference in Atlanta years ago, and he highlighted the power dynamics that often keep nonprofits "begging" for funds or constantly on the defensive to justify their expenses. Pallotta's talk, and his newly published book *Uncharitable*, brought it all home in one simple example he shared from the podium about a tale of two soup kitchens.

> Soup kitchen A reports that 90% of every donation goes to the cause. Soup Kitchen B reports 70%. You should donate to A, right? No-brainer. Unless you actually visited the two and found that the so-called more "efficient" Soup Kitchen A serves rancid soup in a dilapidated building with an unpleasant staff and is closed half the time, while Soup Kitchen B is open 24/7, and has a super-friendly staff that serves nutritious soup in a state-of-the-art facility. Now which looks better? The admin: program ratio would have failed you completely. It betrays your trust. It's utterly deficient in data about which soup kitchen is better at serving soup. It undermines your compassion and insults your contribution. And yet we praise it as a yardstick of morality and trustworthiness. It's the exact opposite.[5]

A similar mistake can be made when evaluating clinics. Just to open the doors at Good Sam we have to spend a lot of money on overhead. Doctors, dentists, psychiatrists, and nurses are expensive. Keeping medications fully stocked and equipment up to date is a major ongoing cost. One new dental chair can cost thousands of dollars, and we have fifteen operatory rooms! Yet every single time a new person enters our facility, they remark that we are not what they pictured when they planned to visit a "charitable clinic." We stand with Dan Pallotta and argue that maybe we should be precisely what they picture.

The overhead myth goes hand in hand with the second trend, which is the growing burden of evaluation placed on nonprofits. While we fully agree that nonprofits should be held accountable for results, we also need relief from the often overwhelming number of different evaluation frameworks and templates required by institutional funders. In addition to the basic letter of intent, application process, and annual reporting schedule, many funders are now requiring project-specific logic models and outcomes tied to every grant. This is a good process and most nonprofits go through these basic planning and evaluation exercises when drafting their strategic plans or setting their annual goals. The challenge arises when, in addition to their established internal processes for measuring success, a nonprofit is required to slice and dice their data in dozens of different ways to match the evaluation requirements of countless funders. It's similar to the problem the university system identified years ago when the majority of colleges and universities moved to the common-application process. The higher education sector realized that they were all essentially asking for the same information from prospective students, but making it administratively difficult for the applicant to apply to multiple schools. A common application and evaluation system in the philanthropic world would be a game changer. Alternatively, funders could allow nonprofits to present their results in the format most useful to them and most helpful for long-term impact tracking in their specific context.

Let us share a real-life example. Historically, Good Sam has produced a monthly "census" in which we report our patient demographics, services rendered, patient payment types, and so on. We also have multiple tracking mechanisms that tie directly to our electronic medical records in order to monitor our quality-assurance measures. These systems have been carefully designed and tweaked over time to reflect the most pertinent information needed by our organization's leadership team to make important decisions and track growth. A few years ago one specific funder requested patient data in such a way that it required rebuilding some of our internal spreadsheets, adding extra steps to our monthly process for internal reporting. It was burdensome, but we made the adjustment so we could comply with the grant and still pull our regular monthly data. Less than two years after this, the same funder completely restructured their online portal and data requirements. It literally required an overhaul of our outcome-tracking mechanisms (again) and ate up an entire week of staff time.

Thankfully Good Sam has the resources (software, staff, expertise) to adjust when needed, but it is expensive and inefficient. For many small organizations or volunteer-led clinics, this would not be financially feasible. In Good Sam's case we have a full-time grant writer and two full-time staff who assist with data analysis, and we maintain expensive subscriptions to house patient information and demographics. In addition we have two full-time staff who assist with marketing and development. Securing large, recurring grants requires significant investment in grant writing and fundraising infrastructure. However, this level of investment is not possible for many small clinics; hence they are unable to compete for the dollars they desperately need. We need more thoughtful funders willing to fund organizational capacity building and operations because of their firm conviction that the healthier the charitable clinic sector is, the healthier our patients will be.

One more note about institutional funding trends. Because dollars are tight and expectations high, many funders are requiring nonprofits to accomplish more within a one to three-year grant term and add new programs and services every year in order to sustain the same (or less) level of funding. Where a clinic may have once received annual funding to maintain national standards for pediatric vaccines, now they may be asked to achieve that success plus additional child health outcomes on the same dollar.

There are also misconceptions at the donor level about how to best help nonprofits. First, donors have to remember that nonprofit professionals are the experts in their field and they know best what is needed to make change. Yet nonprofits are often at the mercy of the donor, whether that donor is an individual, corporation, church, or foundation, and their preferred method of "helping." This plays out in the form of comments like, "We don't want to just write a check, we want to volunteer" or "We're giving to more organizations now to diversify our impact."

The first misconception pits volunteering and giving against each other, as though only one is possible. The reality is, most nonprofits first need financial support and then consistent, skills-based volunteering. The kind of volunteering that drops in for one day to paint a wall, mow a lawn, or cook a meal can also be helpful, but only if it is truly a springboard to more meaningful, long-term involvement.[6]

The second trend, toward donors wanting a more diverse funding portfolio, really doesn't help the nonprofit sector. Each nonprofit needs a core donor family of committed supporters who will stay the course with them through thick and thin. They need people they can count on in a big way, year after year, to sustain their enthusiasm with increased levels of financial support. Personal philanthropy is most helpful to the nonprofit sector when donors pick a few organizations and invest deeply, versus a wide and shallow commitment to numerous agencies. Consider selecting an organization or two who are doing meaningful work in your community and setting up automatic monthly giving. Another strategy is to commit to the percent of income giving in which you give a percent of your income that holds steady as your income increases over time. This is the kind of sacrificial giving we need if we are to adequately address and solve the modern social issues of our day. We are living in serious times, and serious financial investment is required.

If we, as a nation, want to achieve health equity, we must start by recognizing that not everyone has an equal chance to be healthy. People one zip code, one bus stop, or even one street apart can have drastic differences in health and life expectancy due to differences in social determinants. Changing this requires investment in robust, evidence-based program models and sufficient financial support. We have learned during our time at Good Sam that this is not easy. However, the experience of people moving from illness to health is a reminder that the hard work is expensive but necessary and fruitful.

THE FULL CIRCLE OF HEALTH IN ACTION

Gabriella and Mariana were fifteen and sixteen years old respectively when our pediatrician informed them of their positive pregnancy tests. The sisters had received care at Good Sam sporadically over the previous few years and had been brought in by their mother that afternoon. The girls had been young children when they fled Mexico with their mother. They left behind an abusive father to start over far away from his reach. The girls, particularly Mariana, struggled in school having learned English part way through elementary school and coping with the psychological trauma of abuse. As neither was a US citizen, they did not qualify for Medicaid health insurance. Once it was determined that the pregnancies were the results of consensual relationships, our pediatrician referred them to our prenatal program.

At their first visit, their mother asked questions while the girls bickered. We quickly realized how much they did not know about their own bodies and the process of pregnancy and childbirth. The sisters attended our prenatal education group visits, learning about pregnancy, delivery, infant care, and breastfeeding. Mariana started counseling, working through the trauma of her childhood abuse. Our staff encouraged them, brought in baby clothes, and spent hours answering questions. Both sisters delivered healthy babies within a month of each other. The babies received care through our pediatric program, allowing Gabriella and Mariana to continue their postpartum care at Good Sam. Both girls breastfed their babies, Mariana continuing even after her baby's first birthday. Both graduated from high school. For Gabriella and Mariana, prenatal care alone was not enough to adequately address the barriers they faced in completing a healthy pregnancy and entering motherhood. Participation in health education and counseling helped prepare them to be the knowledgeable, caring mothers they have become.

Ms. Swan lives within walking distance of the clinic. She started seeing us for her blood pressure, and during the course of her treatment we discovered she had diabetes. We had a long discussion about her diet and how to quickly control her blood sugar. After that, she didn't return for almost a year. The next time we saw her, she was following up after a hospitalization. She had been hospitalized with an extremely high sugar of over 500 and discharged with little idea of where to begin in terms of managing her diabetes. Usually a gregarious woman with an infectious smile, Ms. Swan looked defeated. She teared up as she told us she was overwhelmed with the idea of diabetes.

We started insulin that day, and she met with the nutritionist during the visit to set some obtainable goals. Ms. Swan also signed up for our diabetes education series and didn't miss a class. Each month, she would see primary care, bringing in her sugar logs for adjustments in her insulin dosing. Her diabetes was improving, but more importantly her smile and commanding personality returned. When she completed the diabetes education series, she started our Eat Well Live Well program, where she attended cooking and nutrition classes. She also started seeing the dentist and a specialist for foot care. Each time she came to Good Sam for a visit or a class, she would leave with a bag of food from the farmer's market. A year later, Ms. Swan reached her glucose goals, successfully controlling her diabetes. She decided to take the diabetes education series again and now can practically teach it. Ms. Swan also gives a short sermon to anyone new who joins the group. "Diabetes is no joke," she tells them. "I almost died. You've got to do the right thing and get your life straight."

Henry was nonverbal the first day we met him. A local pastor had referred him to our homeless program, and we wondered how we would be able to help this young man who wouldn't speak or make eye contact. We slowly learned that he had been living on the streets most of his adult life and had been diagnosed with schizophrenia. Henry was able to see our psychiatrist the same day, and we initiated medication. We had no idea if he understood what we were telling him about the medication or the importance of returning for follow-up.

We were pleasantly surprised when he returned the following week. One of the medical assistants brought him some clothes; he was about the same size as her husband. We adjusted his medication and began to piece together his story. Henry came back each week. He would eat breakfast in the lobby and talk with his team of providers. He was now speaking enough to reveal a speech impediment, and he started making some eye contact. Despite his progress with medication, he was still a vulnerable young man in his early twenties living on the street without a shelter to call his own for as long as he could remember. Even more than medication, this man needed a home. Around this time, we had developed a new partnership with a local social-services agency. The case workers had been bringing patients to us for diagnosis and treatment, and were now starting to complete housing referrals for our patients. Having slowly developed trust with Henry over the last several months, we broached the topic.

"We have a contact that could meet you here next week and complete a housing assessment," we told him one morning. "Working with him will require a detailed interview. They will ask a lot of questions." Henry eagerly agreed and completed the interview, a task that would have been impossible for him just months prior. Even with his improvement, Henry's speech impediment made speaking difficult, and he used his words only when necessary. When we saw Henry a month later, he was thrilled to tell us he was moving into an apartment. This would be the first time he had ever had his own place, and we had never seen him smile so broadly. The case worker also helped him find a lawyer to help him apply for disability, and we were able to complete his medical paperwork. At the end of the visit he grew serious, focusing hard to form each word.

"This . . . this . . . this . . . may sound strange," he told us. "But I want . . . to . . . to . . . thank . . . everyone here for . . . for . . . for helping me. If . . . if . . . if . . . if you need a volunteer, I . . . I am your man." Henry doesn't have a home or even a dollar, but he wants to come and weed the farm or take out the trash because he cares about the clinic and wants to show his appreciation for the help he has received.

For us, this is the kingdom of God. It is Dr. Warren using his affluence to give abundantly. It is Ms. Swan sharing her story with our patients each time she is at the clinic because she does not want anyone else to experience deadly sugar levels. It is Henry, who has almost nothing, giving the only thing he has: his time. When asked why he started Good Sam, Dr. Warren says, "It's just my way of living out my New Testament Christianity. Jesus tells us to do three things—heal the sick, preach the gospel, and cast out demons (Luke 9:1-2). I don't know much about the casting out of demons, but I do know about health care and understand what it means to preach the gospel. I feel like I'm just doing my part to be faithful to my Lord."

Each day at Good Sam we see examples of people who care about the health and well-being of their neighbors. From the dedication of our staff and volunteers to the kindness and generosity of our patients, to the creativity and passion of community partners there is healing work happening in our neighborhood. Creating healthy communities will take time, policy reform, and restructuring of social constructs, but it can be done. It starts with local-level interventions and changing the way we think about our neighbors. The answer to the question, "What will make our neighbors healthy?" is us.

CHAPTER 12

POVERTY

ADDRESSING A DISTANT DISEASE

Mary Kelly grew up poor in rural north Georgia.[1] As the granddaughter of sharecroppers and the daughter of a single mother, her childhood was shaped by the dysfunction of poverty. Her grandparents worked for the Johnsons and were paid in coupons redeemable only at the Johnson Company Store, cementing their lifelong debt to the family. Her Appalachian community was devastated by drug use and currently suffers a higher alcohol and drug overdose and suicide mortality rate than the rest of the United States.[2] Her memories of inadequate educational opportunities and family dysfunction share similarities with the children in Atlanta's poor urban areas.

Now a college professor who has spent her academic career studying poverty, Mary highlights the defining characteristics of poverty in the United States. First, poverty is hypersegregated, meaning it is segregated by both race and socioeconomic status. "So, in Appalachia you have concentrated debilitating poverty, and in the urban environment you see concentrated debilitating poverty," Mary explains. "Appalachia is primarily white and the urban environments predominantly black." Between 2000 and 2012, the number of people living in poverty in distressed neighborhoods like Appalachia and Southwest Atlanta grew by five million.[3] Those living in areas of concentrated poverty face what is called the double burden of not only their own poverty but the impact of those living in poverty around them. Schools, for example, are funded by property taxes, leaving underfunded and understaffed classes in the poorest school districts.

Economic trends in the twentieth century, including the loss of manufacturing jobs, the movement of jobs from cities to the suburbs and overseas, and globalization intensified the clustering of poverty in the United States.[4] Neighborhood disadvantage is then passed on from one generation to the next. As Mary explains, "When you come from poverty and the dysfunction it brings, you do not have the skill set to live in a healthy family." Generational disadvantages fade slowly.[5] US legislation has historically only added to this problem. In his article "Ending Urban Poverty: The Inherited Ghetto," author Patrick Sharkey asks pointedly whether the legislation of the civil rights era, most notably the Fair Housing Act, allowed black children to advance out of America's poorest, most segregated neighborhoods, or whether it reinforced generations of black families to remain there. He says that almost three out of four black families living in today's poorest, most segregated neighborhoods are *the same families* that lived in the ghettos of the 1970s.[6] Poverty is often inherited, undeserved, and self-reinforcing.

However, poverty doesn't have to be a life sentence. Mary is a college professor with a PhD, living in a house she owns in a comfortable neighborhood. She has spent her life in education, living out the belief that with appropriate support structures, people can leave poverty.

CHANGING PERCEPTIONS ON POVERTY

Addressing poverty requires support structures and opportunities for individuals as well as larger-scale reform to prevent the cycles of poverty that are so difficult to escape. This type of change begins with confronting our perception of poverty. In several studies, Americans display a degree of disdain toward the poor. In one study, people were significantly more likely to attribute negative traits to the poor than to the middle class. They were also more likely to name internal attributions, such as laziness or drug abuse, as the cause of poverty compared with external or societal attributions.[7] The same study found that people with a strong belief in the Protestant work ethic held more negative stereotypes of poor people. This comes as no surprise to Mary, who notes that Judeo-Christian heritage has long contributed to the American belief that with enough hard work anyone can make it. If people "reap what they sow," then those with little must be doing little work.

This view is then reflected in our policies as well. As Odetta MacLeish-White, the managing director of TransFormation Alliance explains, "Policy makers often choose one bad story about a person and use that to assume all poor people are that way."[8] As a result, assistance often comes with stipulations. Housing programs require prospective residents to be free of drug and alcohol use before offering an opportunity to get off the street, and obligate mandatory drug screens for food stamp recipients. These are examples of the belief that we should help only the "deserving poor." Access to survival necessities depends on the behavior of those most oppressed, stressed, and vulnerable within our society. As the safety net of support services continues to erode, Americans' expectation that the poor pick themselves up, work hard, and succeed has not waivered. "We expect the poor to succeed in a competitive environment with fewer tools than their middle- and upper-class counterparts," explains Mary. The United States is a society in which housing, education, health care, and even health itself are privileges, not rights.

Poverty, for many Americans, has become a distant disease. With poverty isolated in hypersegregated neighborhoods, we can more easily reinforce negative stereotypes and place the blame of poverty on the victims themselves. Poverty is not alleviated by a single program or strategy. First, we must come to understand poverty as a symptom of systemic injustice. Before living in Atlanta, we had grown up fairly sheltered in a middle-class suburban bubble and had no understanding or appreciation for generational poverty and the impact of structural racism or institutional barriers that exist. We didn't know about the lack of tax base to support schools, the impact of distressed environments, or how the culture of poverty affects how kids grow up and their future. It wasn't until we lived in this environment and started working alongside the poor that we started to see our own biases revealed. Our view of the poor has completely transformed from a negative perception and belief that poverty often ensues because of lack of personal responsibility to a deep compassion and respect for their grit and determination, despite the many daily obstacles they face. In fact, many of the program leaders we interviewed noted the skills and resilience they have observed in those living in poverty. "People [experiencing homelessness] have amazing functional capacity—they are surviving homelessness," explains Sam Tsemberis, founder of Pathways Housing First.[9]

We will be able to confront poverty only when we see the humanity in those afflicted. When we "other" the poor, we have already lost. Changing the perception of poverty begins at an individual level when we choose to address our own biases. Next, we work within our sphere of influence to further societal change. As Cole Thaler, the director of the Safe and Stable Homes Project for the Atlanta Volunteer Lawyers Foundation, works with lawyers who are taking on the cases of low-income tenants, he is engaging in this type of work. "There are persistent myths about who low-income people are. They are on drugs; they are cheats and taking advantage of the system. Part of the work is busting that myth and helping people see that the cause of poverty is almost always the system."

This perception also changes the way we provide assistance for the poor. Many initiatives aimed at helping the poor are flawed from the start when those designing the initiative assume they know what the poor need. In the 1990s, Sam Tsemberis, a psychologist, and his outreach team were working with individuals experiencing homelessness in New York City. They observed the same group of people moving in and out of hospitals and shelters, and back to the street. Tsemberis and his team decided to do something different. As Tsemberis describes it, the paradigm shift in their approach was asking people what they wanted. "We don't know what is best for people," he explains. "We started asking people to set the priorities for their treatment. No one said, 'I want more medicine.' They said, 'If I have a choice, I want a place to live.'" This paradigm shift launched the housing-first movement, a massively successful approach to ending homelessness that we will discuss in detail in chapter sixteen, "A Place to Call Home." However, the success of the initiative did not start with taking people off the street and putting them into apartments. It started with a core belief that people, even those with mental-health illnesses, addictions, and living in extreme poverty, know what they need and are capable of directing their own path to wellness. This is not to say they do not need support. Many will need lifelong supportive services. Effective initiatives match the level and type of support to the expressed needs of the people receiving those services.

In addition to assumptions about the needs of the poor, MacLeish-White points out that we often have judgments about how the poor should react to aid and how successful they should be when given opportunities. Most assistance programs don't offer second chances, because the expectation is success on the first attempt. Gratitude is also a common expectation. Mary remembers three specific families and a few teachers who took her in during her childhood and had a significant impact on her life. However, her childhood was rough, and she did not always know how to respond to the support she was offered. "My behaviors were not always those that showed gratitude," she recounts. Yet these families continued to support her.

We will often have patients thank us for seeing them at the clinic or express gratitude after a routine exam. We find ourselves thinking, *You do not need to thank us for doing our jobs. You deserve the best health care we can provide.*

STRATEGIES FOR EQUITY

When we view poverty as a societal problem, the conversations change from that of what the poor must do to lift themselves out to what constructs of society need to change to bring equity. This change then permeates every aspect of society. King County, encompassing Seattle, Washington, is an example of this type of change. County leaders recognized persistent inequities in health, resources, and opportunities throughout the county due to race and place of residence. County leaders decided to pursue an equitable community, and in 2008, they established the equity and social justice initiative with the aim to "intentionally use an equity lens in departments' policies and decisions, organizational practices and engagement with community."[10] Initially an initiative of the executive branch, in 2010 the county formalized its commitment to equity with the passage of the equity and social justice ordinance. In 2015, the Office of Equity and Social Justice (ESJ) was established with dedicated staffing and resources.

Matias Valenzuela, the director of ESJ, explains that ESJ began operations out of the health department.[11] From the start, however, the initiative has been a thoughtful process aimed at bringing people on board. As Valenzuela explains, "We asked ourselves, 'how do we bring folks not in health into the fold such as people in transportation, parks, and human resources?'" ESJ now works to support all county employees, embedding equity in all county operations. While the ESJ office offers support and training, every county agency and employee shares the responsibility of advancing equity and social justice activities.[12]

This countywide commitment to equity is evident in several innovative government processes. First, the county established measures of equity in order to accurately answer the question, "How do we know if we are progressing toward a fair and just community?"[13] Fourteen determinants of equity guide all operations. These range from quality education to an equitable law and justice system, to access to health and human services. For each determinant, sixty-seven community-level indicators allow the county to measure the extent to which policies and programs advance equity. For example, quality education is measured by third-grade reading proficiency and parent perception of cultural responsiveness among other indicators. Indicators of access to health and human services include unmet medical need, diabetes rates, and tobacco use. Regardless of the department, county programs report which determinants of equity they address, and progress is measured by established indicators. With every new policy decision, budget created, or program initiated, the goal is to move upstream, addressing the causes of inequity rather than only the symptoms. "The goal is to stop the bleeding and ill effects of determinants of health/equity that are not optimal," explains Valenzuela.

Much like Tsemberis asking his patients what they needed to get well, Valenzuela stresses the importance of community engagement. He emphasizes the need to set up a governance team with the community at the table from the very beginning of the decision-making process. During this process, leaders also need to be clear about what the role of the government or leadership is versus community organizations. Equally important is an ongoing commitment to building capacity within the communities. Creating opportunities for high levels of community engagement or community-directed action requires leadership within the community. Building capacity is a step beyond providing aid. "With capacity building," Valenzuela explains, "you're equipping communities to resolve issues for themselves."

Another tangible way equity is introduced into every aspect of county governance is through the equity-impact review. During this process, data is collected to determine the impact of any plan, policy, program, or project on equity, with a particular focus on communities of color, low-income populations, and limited-English-speaking residents.[14] The strategy is also employed during all budgetary processes. The review process asks community members and leaders to consider who the stakeholders are and which determinants of equity are affected by the decision. As Valenzuela describes, this process plays out hundreds of times in small ways creating the expectation that all county decision-making needs to consider equity. He cites their new grading system for food establishments as a recent example. Initially, there was a push to give food establishments a letter grade. However, during the equity review process, they noted significant cultural issues driving the scoring in ways that disadvantaged the international districts. Instead of letter grades, King County is now the first in the nation to use an emoji system that is easy to understand. Each emoji rating is calculated on a curve so that food establishments are compared to establishments in their zip code. This process promotes fairness by correcting for the difference in viewpoints between inspectors. In this process, the equity review changed the course of action.

Assessing equity impact is not unlike asking, What will make my neighbors healthy? King County is moving toward evaluating all decisions through the lens of equity. A similar approach on a larger scale would quickly highlight the ways so many of our national policies promote systems of injustice and oppression. What would it look like if, when evaluating educational policy, we asked, Does this strategy promote equitable access to high-quality education, particularly for communities of color and low-income populations? How would zoning and housing strategies change if we asked, How do we provide adequate and safe housing for everyone in our community, particularly those at the highest risk for homelessness? How might we redesign health care delivery if we asked, Do those most vulnerable to poor health outcomes have access to quality, patient-centered care?

We like the question, What will make my neighbors healthy? for two reasons. First, it requires us to think of the poor and marginalized as our neighbors. After all, our health is tied to theirs. Rather than labeling the poor as other, we accept responsibility, recognizing the ways we are complicit in upholding systems of oppression. Second, the question directs us to the end goal of health. Inequities, whether in education, employment, or socioeconomic status, result in sickness. This sickness robs people of productive lives and longevity while weakening our nation. We cannot make a meaningful impact on health outcomes, whether heart disease, preterm delivery, or diabetes, without addressing the determinants underlying these outcomes. However, with a commitment to uphold the humanity of the poor among us and a willingness to dismantle the systems that enable poverty, we can heal, and we will all be better for it.

CHAPTER 13

HIRED AND HEALTHY

Janet is by all conventional measures on the lowest end of the socioeconomic gradient. She is unemployed and sleeps in a parking garage most nights. She has been homeless as long as we have known her, and she tells us she has already been in and out of all of the shelters. Janet is intelligent and articulate, with a witty sense of humor. Her eyes sparkle when she makes us laugh, but she keeps her deadpan, sarcastic affect like a professional comedian. Janet also has a diagnosis that impedes her ability to form relationships and deal with everyday stressors.

Janet comes to the Good Sam clinic at least twice a month. Despite medication and full participation in her care, living in a society built for the success of the healthy, wealthy, and fully abled has proved challenging at best. Janet's mom repeatedly told her she should work. Seeking disability was a crutch she didn't need. And Janet wants to work. She picks up odd jobs whenever offered and volunteers with her church. A few months back, she was working for an employment-placement agency. She asked for manual labor specifically because "I can lift as much as any man." She followed every policy, and the day she had her first meeting with the agency, she was more excited than we had ever seen her. A week later she returned, and we couldn't wait to hear all about it.

"How did it go? Did they find you a placement?"

"Yeah. . . . They did," she replied, "but . . . they placed me in childcare. They found me a job as a nanny."

We sat in silence for a while. We didn't need to talk about it. We knew that would be disastrous, for her and the children.

With a soft voice, devoid of the usual sarcasm, she finally said, "I lose my temper all the time. I can't take care of a child. I could never be responsible for that. I do not want to hurt anyone. I told them, 'Give me manual labor. Put me in a factory.'"

As a team that has come to love her, we were devastated. As her providers we were grateful that she had the insight and honesty to recognize a bad situation. Not only that but her thought was of the children she had never met. She wasn't willing to jeopardize them for her own gain. The agency didn't see it this way. She told us that they were furious with her. There would be no reassignments or second chances.

Employment is more complicated than simply getting a job. At Good Sam we regularly see the barriers that keep people from gainful work despite their efforts and desire to work. Take Monique, for example. Two years ago we provided her prenatal care during her second pregnancy. She returned a few months ago, late in pregnancy with a baby boy. We rushed to condense nine months of prenatal care into the two months we had remaining. Her husband was abusive, and she fled with her two little girls, leaving everything behind. She made it to a women's shelter designed for women fleeing domestic abuse. Within two months she had obtained full-time employment and moved into her own apartment. On her last visit with us, she sat on the exam table sobbing, her huge belly heaving. She was emotionally and physically exhausted. She will work full-time up until the day of her planned C-section. Her mother has flown in from Africa for a two-month stay. Monique will return to work two weeks after delivery because she has no maternity leave and must feed her family. Her mother will provide childcare for a little while because daycares don't admit two-week-old babies. Monique will work long days, leaving a newborn at home and healing from a major abdominal surgery.

"I'm sorry," she says between sobs. "I'm really fine. I'm just overwhelmed."

Overwhelmed doesn't seem to cover the life of a working single mom of three with her family a continent away. When we hear reports of unemployment or read the debates about minimum wage versus living wage, we think about the stories behind these statistics like Janet's and Monique's. Conversations about unemployment need to focus on the barriers that keep people from gainful, long-term employment. These conversations matter because stable employment makes people healthy. Employment brings a source of income, improved socioeconomic status, and often additional benefits such as health insurance. Yet even apart from its linkage to socioeconomic status, employment is linked to improvements in health. Across numerous developed countries, studies confirm that as employment increases, measures of chronic disease and mortality decrease.[1] When individuals become employed, they experience improvements in mental health and life satisfaction.[2]

Employment as a strategy to improve health is complicated. Even hardworking, motivated individuals face incredible challenges when finding employment, whether due to criminal records, disability, mental illness, language and educational barriers, or needing to finish in time to pick up children from school. In addition, not all jobs are equally effective in promoting good health. Night shift, long hours, working multiple jobs, and jobs with high demand and low control have all been associated with fatigue, job strain, and increases in cardiovascular disease.[3]

Clearly, national policies around issues of minimum wage, maternity leave, and employee protection are essential to substantial change. However, local-level interventions can have an immediate impact. Local interventions also have the ability to focus on the unique needs and barriers that are significant for unemployed individuals, improving health through a multifactorial approach to employment.

One such example is Georgia Works, a privately run program that aims to help chronically homeless men become "productive and self-sufficient individuals."[4] The program seeks to decrease the barriers to employment, connecting men with a range of support services needed to sustain long-term employment and life change.

"Men who come here are really broken," says Phillip Hunter, the executive director of Georgia Works who launched the organization in 2013 with three other employees.[5] "They come from things I can't even imagine." For six to twelve months men reside in a sober living community where they receive education, counseling, employment, and support services to meet their unique needs. During the first thirty days, men receive a case worker and develop a treatment plan much like the kind we make in the clinic with patients, except this treatment plan includes tasks like obtaining a GED, securing a driver's license, paying off bad debt, and mending broken relationships. While Georgia Works is providing social and psychological support for participants, it is also a staffing agency.

Phillip Hunter explains that the labor market is ripe with jobs. However, men entering the program have histories of prolonged unemployment, homelessness, and often substance abuse, creating reluctance among employers. This is where Georgia Works steps in. "We tell employers we'll do your HR and weekly drug screens," explains Hunter. The situation becomes a win for employers who obtain reliable, affordable temporary employees, and for program participants who receive minimum wage. After three or four months these men are hired as full-time employees who then receive a living wage.

Georgia Works also specifically addresses mental health. Don Miller, the full-time counselor, explains why many of the men have dual diagnoses. Drugs are self-medication for unresolved trauma. For many, this is the first opportunity they have had to address these issues. Addressing trauma and behavioral issues are key components of the program. Men receive counseling and can be networked to long-term therapy as needed. "We change the question," Miller tells me, "from 'What did you do?' to 'What happened to you?'" Hunter explains that behavioral health and medical issues don't always present until a participant is working. "We can't operate in silos," he explains. The program leadership and staff understand the intertwined relationship between health and employment.

Chandler, now a case worker at Georgia Works, found Georgia Works at a turning point in his life. "I decided to change the way I was living," he explains. "No one would give me a job, and I was trying." He finally found work, four hours two times a week and lost it because the employer didn't have time to train him. "I wanted to change the way I felt inside about myself, not just get a job," he remembers. "My goal was to be able to help someone." Chandler found his chance of sustainable employment at Georgia Works and later became a full-time case worker for the program. "I'm a Georgia Works man," he says proudly. Chandler's story is shared by many participants as 80 to 90 percent receive full-time employment upon completion of the program.

The Georgia Works program has also proven sustainable, costing $2,500 per participant.[6] This pales in comparison to the cost of incarceration, $21,000 annually per inmate in Georgia, or the cost of caring for someone who is homeless.[7] A cost-benefit analysis factoring in incarceration costs and loss of earning found that for every dollar invested in Georgia Works, the program returned between $6.10 and $11.30 in individual and community benefits.[8] The program saves costs for both the judicial system and health care system, and is now in the initial stages of replication in two additional states and another site in Georgia.[9]

This comprehensive approach to ending homelessness recognizes that sustainable employment requires social, behavioral health, and medical support systems. Participants leave with more than a job. They leave with improvements in health and life expectancy. "We give them something positive to identify with," Hunter explains. "They leave saying, 'I'm a Georgia Works man.'"

As a society, we like stories like Chandler's, where people recover and succeed, but initial success is not always the reality, highlighting the need for programs that address the individual needs of people seeking employment and offer second, third, and fourth chances. James was initially an employment success story. He arrived at Good Sam a month after a relapse had rendered him homeless. He heard about the clinic at a recovery meeting and told us he was starting over.

"I'm getting too old to keep doing this," he told us. "I'm here because this is the last time. I am going to turn my life around."

He did. He started attending counseling, primary care, and psychiatry visits every other week and then monthly. He participated in integrated care, working with a team of providers to discuss and treat some of the underlying conditions that had fueled his downward spiral into relapse. He received housing at a men's shelter and an assigned caseworker helped him establish goals. He started working in a kitchen. James had no particular interest in cooking, but he worked hard and generally enjoyed working every day. His hard work earned him a job in the private sector in a sales position where he had previous experience. He excelled and his temporary contract became a full-time position.

He announced his new position with pride and shared his plans for a better life. During our visit that day, we talked about examining benefits packages and enrolling in a health insurance plan. We discussed how health care works when you have health insurance. We talked about maintaining follow-up, medications, and recovery meetings. This was a start of a new era for him, but he understood it did not erase the past.

We called to check on him a couple months later when he didn't come for follow-up. He didn't respond, and we hoped it was due to his busy work schedule. Then, a few weeks later he sent us a message: "I need to come back. I'm back in the shelter." Two days later he was meeting with our psychiatrist again and starting back on his medications. He had relapsed, but this time it didn't end in imprisonment or hospitalization. It had cost him his job, but he sought help. He returned to the case worker and shelter he knew, with the confidence that he could succeed again. He called us because he understood that his addiction couldn't be overcome without support. While our hearts ached that he was starting over again, this is also a success story. He trusted us as his health care providers to welcome him back without question or judgment. Today, he is back in full-time employment.

Everyone has benefited from second chances. Whether our mistakes were relational, financial, or spiritual, we have all experienced forgiveness and new starts throughout life. Creating pathways to sustainable employment and health requires space for relapse and healing. As Odetta MacLeish-White, the managing director of TransFormation Alliance, asks, "Why do we expect poor people to be perfect the first time they are given an opportunity?" Programs like Georgia Works accept brokenness at face value. The success of Georgia Works and programs like it lies in the ability to surround people with unrelenting support. "Everyone has different issues," says Hunter. "Whatever it takes to get them through is what we do."

Individuals can impact the availability of employment opportunities and support others in establishing gainful careers. When we asked Phil Hunter how individuals could support his organization, his answer was surprisingly simple. He explained that he would love a group of consistent volunteers to teach computer literacy. If you have a skill, be it computer literacy, résumé building, or cooking, find a place to teach that skill. If you are an employer or manager, consider how you are equipping your employees for career advancement. Can you offer an employee-development plan, educational opportunities, skill-building sessions, or financial-management training for your employees?

We launched an employee-development program at Good Sam, recognizing that some employees will outgrow the positions we have available. We may lose good employees, but increased career mobility is good for their health. As an employer, are there positions within your organization where you can hire an individual leaving prison or rehab? For Chandler and James, someone said yes, and that answer changed their lives. When we have opportunities to be that *yes*, we can be a part of improving someone's health and longevity.

CHAPTER 14

GROWING HEALTH FROM THE GROUND UP

As the sun rises over the Atlanta skyline, Good Sam's lead farmer, Cody, pulls supplies from the farm shed. We exchange a "good morning" as we head into the clinic and Cody to the field. Our day will be spent indoors: for Breanna, questioning and examining patients, and for Veronica, touring potential donors and submitting grant applications. Cody plans, plants, weeds, and harvests in the hot Atlanta sun, supplying our farmer's market with fresh produce. A farmer is not a typical member of the health care team, and yet Cody's work has the potential to improve the lives of our patients in ways prescriptions never could. Together we are providing an innovative "Farm to Health Care" system for our community.

The farm at Good Sam stands out as a green oasis in a stretch of concrete. Located in the middle of a designated "food desert," patients pass corner shops and gas stations on the way to the clinic from every direction. The one-acre farm, complete with a green house and pecan trees, produces over thirteen thousand pounds of certified naturally grown produce each year, all of which are used for Good Sam's educational programs and sold directly to patients and community members at our daily farmer's market. The farm is a solution to the health conditions we are treating inside. Today, over a third of adults and a sixth of children in the United States are obese, and obesity-related illnesses such as heart disease and diabetes top the list in causes of death.[1] Obesity is more prevalent among low-income communities, placing low-income individuals at an increased risk for obesity-related illnesses. Counties with poverty rates over 35 percent have the highest obesity rates when compared with wealthy counties.[2] Segregated neighborhoods with high-cost and low-quality grocery items have poorer nutrition.[3] Reversing this trend requires access to affordable nutritious food and the knowledge of how to prepare it.

The idea of prescribing food and using food as medicine is not new. Across the United States, health care systems are incorporating new strategies to impact the diet of their patients. From tours of the grocery store to cooking classes, the medical world is recognizing that the prevention of chronic disease starts outside of the exam room. Good Sam included health education in its earliest days. Shortly after its opening, Good Sam hired a nutritionist to meet with patients and discuss their food choices and nutrition goals. However, simply telling someone they should eat better rarely creates change. With the clinic located in a food desert, Good Sam decided to do something about food access. In 2013, Good Sam broke ground on its urban farm with the vision of creating a sustainable food source for the community. Today the farm thrives, producing fresh produce almost year-round. Patients attend cooking classes, practice preparing meals, and purchase the produce on the way out the door from their office visit.

Even once accessible, nutritious food must be affordable. Good Sam has a farmer's market set up at the front of the clinic most days of the week, depending on the season. Patients and community members can purchase produce grown on the farm for low prices. Through funding from Wholesome Wave Georgia, food stamp recipients are able to double the value of their SNAP benefits. For each dollar they spend at the farmer's market, they can receive two dollars' worth of produce.

Once nutritious food is accessible and affordable, people need to learn how to prepare and incorporate produce into their diet. In 2010, Good Sam opened its teaching kitchen, a commercial-grade space on the second floor of the building, and provided a program in which patients could watch chef-led demonstrations and gain hands-on experience preparing new foods. In 2016, Good Sam launched Eat Well Live Well (EWLW), a family-friendly food prescription program in partnership with Open Hand Atlanta and Wholesome Wave Georgia. Providers can sign up patients for the program, and they can enroll their friends, family, and neighbors. Participants in the six-month program attend weekly nutrition classes that include a cooking demonstration in the teaching kitchen. They cook, sample foods, and take home recipes to try on their own. As they leave, they receive a food voucher to obtain free produce at our farmer's market, as well as a bag full of healthy goodies and cooking tools to build their kitchen supply.

Debbie wasn't a patient at Good Sam when she learned about EWLW at a community health fair. She signed up along with her husband hoping to spur them to make some nutritional changes. She had attended other nutrition classes previously. "Normally you leave a class thinking, *That was nice*, but you don't really change anything," she explains. EWLW was different because she was offered a hands-on experience. "Knowing you can go home and do what you just learned is the most important part of EWLW," she explains. She laughs and adds, "There's no excuse to not make changes when you have the veggies to bring home!" Debbie tells me that she and her husband cook at home now. They plan and prep meals and read nutrition labels more critically. They even have a raised garden bed in the backyard. "We're enjoying our meals better now that we know we are eating good food."

Part of the magic of the cooking classes is that they take traditional foods our patients love, like fried catfish or Thanksgiving dressing, and remake them in a healthier way. We're not asking people to change their lifelong palate, just make a few adjustments to improve the nutritional impact of their favorite meals.

More recently, Good Sam's Community Partnerships Manager, Anthony Wilkes, launched Eat Well Live Well for Kids. Anthony started at Good Sam managing the pop-up farmer's market. He lives in Good Sam's neighborhood and his passion for his community and ability to make healthy eating fun was quickly evident. A year into his time at Good Sam, Anthony became the Community Partnerships Manager, with the responsibility of helping Good Sam connect with our community. Anthony attends community events, parent-teacher meetings, local neighborhood meetings, and health fairs almost daily.

While he helps people in the neighborhood connect with Good Sam for care, even more importantly, he educates all of us on the needs and wants of our neighbors. He helps ensure our programming is relevant to our community. We love walking past the large glass windows and peering into the kitchen during Anthony's Eat Wall Live Well for Kids classes. Teenage boys, sporting aprons, are preparing French toast with fresh fruit alongside a local chef and learning to plate and present their meals. We hope they return home and help their parents in the kitchen. Someday, they'll be fathers, and we imagine them teaching their children the value of understanding food and preparing healthy options at home, serving up meals on a nicely dressed table.

Prescription medication plays an important role in chronic disease management, but medical care is limited in its ability to improve nutrition-related illness. As Debbie points out, "Someone is much more likely to eat every day than take a medication every day." Providers can increase insulin and add medication to manage nutrition-related illness, but these efforts are buying time. Long-term change in health outcomes occurs out front at the farmer's market and upstairs in the teaching kitchen. This is exactly why clinics need farmers and chefs.

This intersection of food access and education can also be a solution to the mothers in Veronica's neighborhood filling bottles with Kool-Aid. Our prenatal program, Healthy Pregnancy, Healthy Childbirth, Healthy Parenting, is designed to increase mothers' knowledge and self-efficacy.[4] During the course of their prenatal care, women are offered the opportunity to attend three group visits that focus on education and discussion. During the first trimester we discuss pregnancy with a focus on nutrition. In the second trimester we review labor and childbirth. In the third trimester, we address parenting, spending most of the time on infant feeding and breastfeeding. Women also leave with a voucher to spend at the farmer's market. Our goal is to encourage mothers to believe they can impact their health and the health of their child through good nutrition. Stories like Gabriella and Mariana's are reminders that young mothers from difficult social and economic conditions can make healthy choices. The chubby cheeks and steady growth charts on their toddlers are evidence that those choices have a lifelong impact.

While Good Sam stands as a great example of food access at the clinical level, it takes an entire city working together to truly provide year-round, local, sustainably sourced food for all. There is a powerful movement underway in Atlanta that aims to do just that: create a livable, resilient Metro Atlanta, where everyone is growing, sharing, and eating healthy, local food. The Food Well Alliance, chaired by Atlanta Food Bank founder Bill Bolling, has made this their mission, and they are engaging every aspect of Atlanta's local food system—producers, processors, distributors, consumers, and composters—to make a difference. In their 2017 Baseline Report, they highlight findings from their first few years of research and advocacy: "We've learned that food associated with high levels of well-being, of social justice, of stewardship, and of system resilience is grown locally by farmers and gardeners right here in our city. We learned that both teachers and health care providers are critical to awakening Atlantans about the importance of eating local, fresh, nutritious food."[5] Specifically they have found that equitable access to local, sustainably grown food results in four specific benefits: community vitality (neighborhoods coalescing around growing and enjoying local food), environmental stewardship (building a local food system that protects and improves the urban landscape), health and nutrition (raising the public's awareness of how to grow, cook, and eat nutritious meals), and economic development (restaurants sourcing local food).

But even if your city doesn't have a robust local food movement, there are still important ways you can make your voice heard through your choices as a food consumer. Whether you're ready to start or join a local food movement in your city, the Food Well Alliance suggests simple, tangible ways to increase the availability of locally grown produce.[6] First, spend your money on local food and encourage the places you shop to purchase locally grown food. Purchasing fruits, vegetables, and other food products directly from urban farmers and local farmer's markets or joining a farmer's community-supported agriculture program promotes the sustainability of local markets in addition to improving the quality of your food. Similarly, you can commit to eating at restaurants that make direct relationships with local farmers a core practice of their business. Finally, start your own garden. Like Debbie, growing your own food allows you to reconnect with where food comes from and increases your access to healthy food. Your support of local farmers, community gardens, and neighborhood farmers' markets creates a sustainable and accessible supply of healthy food within the community.

Anthony Wilkes explains it this way: "Like a good recipe, a good community has to have the right ingredients: strong local schools, community advocacy, parks and recreation, enrichment agencies, and of course quality grocery stores and food establishments coupled with health education. When these assets are present, it creates a neighborhood banquet that everyone can enjoy."[7]

CHAPTER 15

A HEALTHY START

It is easier to build strong children than to repair broken men.

FREDERICK DOUGLASS

As a neighborhood clinic serving patients of all ages, we have the privilege of watching kids grow up. We learn about their families, schools, interests, and hopes for the future. We also witness firsthand the relationship between health and education. One particularly memorable teen is Kelly, who had Medicaid and lived down the street from the clinic, making it easy for her to come in anytime she had a concern. Kelly was a motivated student who lived with her mother, older sister, and much younger sister in a small rental house. Her parents emigrated from Mexico shortly before her birth, and her father abandoned the family after the birth of her younger sister. At fifteen, she was comfortable running her household and quickly sized up the newest provider. Breanna recalls their first meeting:

"How old are you?" she asked me. At twenty-four I was aware that upon learning my age most patients would quickly leave in search of a more qualified health care provider. I had a whole list of responses to kindly escape that question. However, I had made a habit of telling my teenage patients the truth. Maybe it was because at sixteen, twenty-four seemed pretty old. Or maybe, I thought, it would show them that you don't have to be all that old to do what you want with your life.

"I'm twenty-four," I answered. Kelly proceeded with a list of questions about how many years it had taken me to earn my various degrees. After her friendly interrogation, she smiled. "I'm going to be the first one in my family to go to college," she told me confidently. "That's awesome," I replied. "You should definitely do that."

Kelly was a frequent visitor to the clinic both for her own health concerns and to bring her younger sister in for care. As minors, the girls needed their mother's consent for most treatment, which proved difficult. Their mother was a sweet, soft-spoken woman with limited English. She worked two full-time jobs to provide her daughters with a comfortable home in a decent school district and was rarely able to get them to the clinic around her work schedule. We helped her write a consent saying the girls could be seen by us anytime for any reason, and she happily signed. Kelly and her older sister raised their younger sister while their mother provided for the family. Her little sister ended up in my husband, Matt's, first grade class. During parent-teacher conferences, Kelly and her older sister showed up because their mother was at work. In a school where only three parents showed up for conferences out of a class of twenty-four students, Matt decided to hold the conference with Kelly.

Over the next couple of years, I got to know Kelly well. I did her sports physical for track, and we talked about running. She told me about her boyfriend, and we talked about safe sex and contraception. She always talked about school and her plans for college. During her senior year she asked me to write her a recommendation letter for college. At first, I was surprised. I knew little about her academic performance and had never even read a sample of her writing. Yet, as I thought about what to write, I realized I did know that Kelly was the type of person who made sure her younger sister never missed a school conference or checkup. She was the teenager who managed her health care, schoolwork, and college applications without the assistance of a single family member. She was the student who could manage her work and do the grocery shopping for the family and put her sister to bed each night. I knew she could succeed at being the first in her family to go to college, and her little sister would have someone to follow.

Kelly didn't get in to her first choice of college. She was devastated that week, but then the offers started coming in. She chose a good state school and majored in business and accounting. Matt and I attended her graduation party with pretty much the entire neighborhood. She wore a white dress, and her mom beamed while filling plates of food. We met her boyfriend, also an accounting major, and they talked excitedly about their plans for employment and buying a house.

As a health care team, we did very little in the way of inspiring Kelly's success story. Her own determination and motivation propelled her from an excited young teenager to a college graduate with a newly launched career. Kids growing up in poverty have what Dominique Jordan Turner, CEO of Chicago Scholars, calls "invisible talent" or "superpowers" gleaned from the resilience required just to survive in impoverished neighborhoods. Growing up in poverty results in stress and brain trauma that can lead to poor health outcomes throughout life. However, the experience can also develop a level of grit, perseverance, and determination.[1]

According to Turner, the skills developed by many young people who grow up in poverty are creativity, collaboration, communication skills, critical thinking, and people management (or life management with an adult perspective at a young age). Interestingly, Turner says these are the same skills she looks for when hiring senior leaders for her team. This does not mean that it is acceptable or desirable that over 20 percent of US children live in poverty.[2] It is a reminder that kids like Kelly and the youth in Veronica's neighborhood are uniquely equipped to be leaders and productive members of society. We all lose when young people are denied pathways and support for gainful employment and continuing education. Turner argues that youth living in poverty need a mentor who can be a "poverty translator" and point out to them where they are uniquely skilled and have the potential for greatness. Investing in literacy, educational equity, and opportunity for advanced education is perhaps our best hope for change.

Our experience with Kelly demonstrates the interplay between health and education. The public school where Matt started his teaching career had a nurse, but she was shared between two other schools. Even when she was at the school, telling kids they really needed to see their doctor did little if their parents couldn't get them there. Matt had to learn how to use an asthma inhaler in case the nurse wasn't there and one of the kids had an attack. Matt used to suggest parents swing by the clinic when he had to make awkward calls about the growing rash he'd noticed on someone's arm or pink eye that didn't require a degree to diagnose. One day, his principal called the clinic.

"I have this little guy here who cut his eye in P.E. three days ago. It's looking worse, and his mom still hasn't gotten it checked out," she told me. "Can I send him to you all right after school when his mom picks him up? Just tell me what it costs, and I'll send home the payment with Matt." We did just that.

When considering education as a social determinant of health, health care providers can do more than help sick kids get back to school quickly. If we are going to solve education disparities, people other than educators are going to have to become part of the solution in reaching at-risk kids. As discussed in part one, waiting until a child enters the school system in kindergarten is often already too late to change their educational trajectory. Fortunately, innovative strategies are already occurring before children reach the classroom.

When a newborn comes to Good Sam for the first time, we celebrate. Tre, the energetic baby-loving grandmother who grew up in Atlanta, abandons her duties as a medical assistant to meet the new little one. After doting over the baby for a bit, if the parents are willing, she will parade the baby down the hall, making sure everyone in the building has seen the "cutest baby ever." Often these babies belong to moms who have come through our prenatal care program, and we become more like proud parents than health care providers for a few minutes. Although we are celebrating, we are also keenly aware of the complete vulnerability of these infants, and the awesome responsibility their parents now hold. The medical team at Good Sam may be the only professional support this family receives for the first five years of this child's life. Talk With Me Baby was born out of this reality.

Children learn language from their parents. Language learning may be supplemented by a sitter or early child-development program, but even in these cases parents are the first and best literacy instructors. Babies, however, do not receive the same language exposure across all socioeconomic statuses. Children in low-income families hear approximately six hundred words per hour compared to children from higher-income families who hear approximately two thousand words per hour.[3] This results in a thirty million word gap by the age of three and significant gaps in reading ability by third grade.[4] This pivotal study prompted a surge in programs throughout the United States aimed at improving literacy by closing the word gap. Georgia's Talk With Me Baby is the first statewide program. The goal of the program is to encourage parents to talk with their babies.

Given the early and consistent contact health care providers have with newborns and their parents, Talk With Me Baby trains nurses and other health care providers to demonstrate the importance of parents talking with their baby. In the clinic this means we are telling parents that talking, reading, and singing with their baby improves development and literacy even though the baby may not appear to comprehend language. Next, we demonstrate how to do this since having a prolonged, one-sided conversation feels awkward for most new parents. Amy Becklenberg, a nurse practitioner who helped implement the program in an Atlanta refugee community, notes the simplicity as the strength of the program.[5] For health care providers and parents, "This doesn't have to be anything additional," she explains. "Do it when you are doing what you already do. Narrate what you are doing."

In the exam room this means that after a provider greets the parents, she turns to the baby: "Hi Baby! Welcome to the world! It looks like you are doing a good job eating your milk and getting big. Did Mommy or Daddy pick out that cute outfit? I'm so honored you dressed up to see me today!" Then she turns back to the parents and says something like, "Did you know that the more you talk, read, and sing to your baby, the more connections are made in the child's brain and the easier it will be for the child to learn the language? One easy way to help your baby is to talk to your child all day long." With a simple approach, providers are beginning the conversation about literacy and the importance of words long before that baby will enter the school system.

The research leading to the development of programs like Talk With Me Baby is not without criticism. Hart and Risley's study may overestimate the word gap. The highly educated families in the study were primarily white and may have become more talkative in the presence of researchers, while the less educated, primarily black families may have withdrawn, fearing judgment from the researchers.[6] Nevertheless, other studies have demonstrated that vocabulary and language gaps exist between children of lower and higher socioeconomic status.[7] Ms. Becklenberg acknowledges that many low-income families do provide lots of language nutrition. The benefit of Talk With Me Baby is that it can be applied universally, catching families of lower or higher socioeconomic status where language nutrition is not happening enough.

Perhaps the more concerning criticism of word-gap research is the way focus is directed toward individual dysfunction and failure, placing the blame on parents rather than systemic inequities that have a more significant impact on early childhood development and learning.[8]

Direct talk strategies like Talk With Me Baby have the benefit of being straightforward, accessible, and able to be implemented in a variety of settings. Such strategies do not substitute for higher-level education and social-policy changes that address the societal structures limiting low-income children. As explained by the American Anthropological Association, direct talk programs must occur alongside economic and health care changes to produce academic success.[9]

Health care providers can also promote educational success with early identification and early intervention. Dr. Jen had been volunteering as a pediatrician at Good Sam when her son was diagnosed with autism. Her personal journey created a desire to help the families of Good Sam have access to the care that was essential to her son. She connected with Dr. G, a developmental pediatrician. Developmental disorders are common, with one in six children having some type of developmental delay or disorder.[10] As Dr. Jen describes, children living in poverty are at greater risk for developmental delays and greater risk of not being identified and receiving effective interventions. Together Dr. Jen and Dr. G assembled a team, including a psychologist, speech therapist, and social worker, launching the Good Sam Developmental Clinic. "We do not see large numbers," explains Dr. Jen, "but we do family-centered care. We ask families what concerns them." The program is designed to support families with children who have developmental disorders by prioritizing the needs of the family. The team isn't prescribing medication or even conducting therapy sessions. Rather, they are addressing the barriers that prevent families from maximizing their child's ability to succeed. "My role is family support and social worker and secretary," Dr. Jen says with a laugh.

Dr. Jen recounts her experience with a family whose young son had autism. Both parents worked, and the mother, Lorrie, was exhausted trying to balance his need for therapy and work. Lorrie explained that her job offered her the option of three longer shifts per week versus her current schedule of five days per week. She was unable to take this opportunity because she had to drive her son to school each day. Dr. Jen further explored this barrier and learned that her son could not take the bus because he was unable to wait at the bus stop. Dr. Jen called another patient's mom who worked as a bus driver, who was able to provide Dr. Jen with options for the family. Dr. Jen then called the bus company and was able to arrange for the bus to pick up the young boy at his house. Lorrie was able to change her work schedule, reducing her stress and giving her the flexibility to make her son's many needed appointments without missing work. "The bus issue would not have come up if we only asked our routine questions about things we thought were important and didn't ask what successes and what difficulties this family was experiencing," Dr. Jen explained.

The limitation of the developmental-clinic model is that it is time intensive, requiring one to two hours with each family. Most clinics cannot sustain this type of model and remain financially stable. While effective, this intervention does not generate revenue. Increasing the reimbursement and funding for this social determinant-focused care for children with developmental disabilities is the key to its growth and expansion. Many children with developmental delays, particularly those of lower socioeconomic status, will not have access to early identification and intervention before entering school. Changing educational disparities requires interventions that begin well before a child even enters the school system.

All of us can support early childhood education and interventions by advocating for the funding of these programs. When cuts are proposed for public education or early childhood-development programs, call your representatives and request full funding. Consider asking your local school what supplies they need that are not covered in their budget. Maybe they need snacks for kids who come to school without enough to eat. Perhaps they could use more books for their classroom libraries. Listen to educators who work with these children daily and donate according to their needs. Finally, mentorship and tutoring can make a significant impact in the life of someone like Kelly, who became the first in her family to go to college, or Treveon, who lost his life before finishing high school. Volunteer at a local kid's club or afterschool program. Build connections, provide encouragement, and talk about college or career paths. Early interventions are critical, but every action we take to promote and prolong the education of our youth increases their likelihood of longer, healthier lives.

CHAPTER 16

A PLACE TO CALL HOME

Darrell, who was referred to Good Sam by his pastor, lives with bipolar disorder and severe difficulties with anger management. He was so agitated when he first came to the clinic that he almost walked out of the initial exam. He wanted medication, and he wanted to feel better. He had spent the last five and a half years living on the street, and keeping track of his medication was nearly impossible. One week, a water bottle leaked in his bag ruining the medications he was keeping in a plastic bag. Another week, his backpack was stolen. After a few months we learned to have him seen as soon as he arrived. If left alone in the lobby too long, he would begin shouting, and we worried physical violence would be next. Our counselor and psychiatrist met with him regularly, but our efforts were limited by his inability to stay on medication. "He just needs to get off the streets," our psychiatrist lamented.

We referred Darrell to our social-work partners, and we completed paperwork verifying the severe and persistent nature of his mental illness. He was able to obtain housing in a nearby affordable housing community. He returned a week later for medication. He sat quietly in the lobby and smiled, saying "good morning" to the staff on his way back to the exam room. His change in persona was astounding.

"My room has a bed, a desk, a chair, and my own sink," he exclaimed. "After five and a half years on the street, I feel like I am living in a mansion!" His medications were organized in their bottles, and we talked about starting treatment for his chronic hepatitis C, something that just a few short weeks ago seemed impossible.

"It's hard to believe he is the same person," our psychiatrist said after seeing him that day. While our medical care and the medications we dispensed have been helpful, they did not account for this dramatic change. Having a bed at night, a safe place to keep his things, and the quietness of a private room was the therapy Darrell had desperately needed for years.

Others share Darrell's experience. In our work at Good Sam, we often find that homelessness most potently illustrates the impact of social determinants on health status. Good Sam's homeless clinic launched in 2015. Each Friday patients receive primary care integrated with behavioral health as well as dental care, nutritional consultation, and medication. Our Friday clinic embodies the best of comprehensive, patient-centered care. Yet the best of clinical practices cannot mitigate the health impact of homelessness. From the earliest days of the clinic, people would ask, "You can't get me into housing, can you?" They would gladly accept health care and medication, but what they desired most was something our health care providers couldn't give: shelter to call their own. We found ourselves thinking, and often still do, *I wish I could write a prescription for a home.*

The persistent correlation between housing and health we see at Good Sam would come as no surprise to proponents of Housing First like Sam Tsemberis. Tsemberis, a psychologist, was working as the director of New York City's emergency outreach team for the homeless when he and his team initiated a paradigm shift in strategy to address homelessness. His clients, like many of the people who come to Good Sam on Fridays, were in and out of hospitals and not getting any closer to stable housing. Clearly, traditional housing programs weren't working for many of these people. Traditional housing programs work in a step-like approach. Housing starts with emergency shelter and progresses toward subsidized independent living environments. Often, each step requires more preconditions such as medication adherence or abstinence from substance abuse.

But what about Darrell, for whom crowded shelters provoked paranoia and anger? When we first met Darrell, he told us it was safer for him to sleep on the streets than spend a night in the shelter. Shelters do not provide a place of safety, healing, and recovery for many of our patients. After listening to the desires of his clients, Tsemberis founded Pathways Housing First in 1992, a nonprofit organization where he could test the assumption that anyone experiencing homelessness can achieve stability if provided with the appropriate support.[1] The Pathways Housing First program is based on the belief that housing is a basic human right, and therefore individuals with complex clinical needs are placed directly into independent housing without having to prove they are worthy of being housed. They do not have to participate in psychiatric treatment, attain a period of sobriety, or any other preconditions to prove they are ready for housing. Once housed, they are offered the support they need to maintain independent living, which includes meeting the terms and conditions of a standard lease and paying 30 percent of their income toward their rent.

The Pathways Housing First program is a clinical and housing intervention composed of three components.[2] First, the program is based on the philosophy of consumer choice. "We ask people to set the priorities for their treatment," Tsemberis explains. "It involves a shift in clinical practice from traditional medicine to a client-centered, psychiatric rehabilitation approach." The clients decide where they live and what support they need. This leads to the second component: community-based mobile support services. Tsemberis stresses that Housing First is not simply a housing program. "Service support must be at the level of intensity as [clients'] clinical needs," he explains. Clients receive home visits to support them in their treatment plan, which might include sobriety or psychiatric stabilization. This support is also ongoing, changing to meet the current needs of the client. As Tsemberis states, "People get healthier and better much quicker than they get richer." The final program component is permanent scatter-site housing. Clients live in neighborhoods of their choosing and are not housed within designated, segregated areas.

The idea of providing people with independent housing regardless of sobriety or mental status was risky. Even Tsemberis and his team had their doubts. But within months of the initiation of Housing First, the results were clear. Tsemberis recounts his first report to the board of directors: "We had an 84 percent housing retention rate for the first fifty people we housed." The next step involved conducting a randomized-control clinical trial called the New York Housing Study, where 225 individuals with complex needs were randomly assigned to either the Pathways Housing First program or treatment as usual. Two years into the study, Housing First participants were less likely to spend time homeless and less likely to be hospitalized for psychiatric illnesses than those in traditional housing programs.[3] In addition, Housing First participants were stably housed 80 percent of the time compared to 30 percent for those in traditional programs.[4]

As Housing First expanded to other cities, additional data mounted with evidence continuing to demonstrate the significantly more effective outcomes of this client-driven approach to housing and services for individuals who were homeless and had complex clinical needs. The research also showed that there were significant cost offsets associated with this approach because people housed with supports were less likely to use emergency rooms, inpatient services, jail days, detox, and other expensive, acute care services, proving the effectiveness of immediate housing access. The research was conclusive that "providing immediate [housing] access to someone with mental illness and addiction was not only humane but cost effective."[5]

What would it take for Housing First to end US homelessness? Tsemberis's answer is simple, "Resources and political will." More funding for Housing First would mean fewer shelters and more stories like Darrell's. "Homelessness is a symptom of the financial and social policies that drive capitalism and continue to create increasing income disparities," Tsemberis explains. When the housing market is unregulated and rent costs are driven by market forces rather than a government policy that seeks to provide decent, affordable housing for all, the people who are poor or living on fixed incomes are quickly priced out of the rental market and end up homeless. "Homelessness is the tip of a much larger social economic problem," he concludes. Yet it is not an unsolvable problem. Change starts by telling the stories of people like Darrell with supplemental data to prove the cost savings of housing the homeless. Change begins when we believe we can do better, and Housing First is proof that we can.

Good Sam's Friday clinic is dedicated to providing care to people experiencing homelessness. As a result, providers are always asking patients about housing and considering how we might effectively refer and collaborate to facilitate housing. It can be easy to forget that many of the patients we see Monday through Thursday are not far from homelessness themselves. They don't ask us about housing because they are seeing us for medical care. However, we are learning the two are intertwined. An asthmatic patient does not seem to be improving on multiple inhalers because he is living in a rental home plagued with toxic mold. A pregnant patient is hesitant to answer questions about domestic abuse because leaving the abuse will result in homelessness. We have other patients "doubling up," meaning a friend's couch or a relative's guest room is keeping them from homelessness. They have no permanent residence, and there is no guarantee the bed they sleep in tonight will be available tomorrow. Other patients space out visits and skip free mammograms because one less day of work can mean eviction. These stories are commonplace across the United States and, as providers, if we aren't asking about housing, we might miss the single most important factor affecting our patients' health. People moving out of poor neighborhoods and into wealthier ones have shown improvements in mental health, obesity rates, psychological distress, and depression.[6]

In school, providers are taught not to miss red flags when diagnosing patients. In other words, don't prescribe painkillers for a headache and miss a stroke, confuse a mild heart attack for indigestion, or ignore sudden weight loss as a potential sign of cancer. If we are hoping to improve the health of patients, our questions are often already missing the red flag. We should be asking where our patients are sleeping. Is it safe and secure? Is it permanent? Is it contaminated with mold, or does it lack electricity? Is it close to transportation or places to buy fresh food?

Addressing the impact of housing and environment on health goes beyond eliminating homelessness. We also need to consider the factors that lead to homelessness and the conditions that trap people in unhealthy environments. As Cole Thaler of the Atlanta Volunteer Lawyers Foundation explains, solutions range from microlevel to federal interventions, and advocacy is needed at all levels.

At a local level, city ordinances can impact the availability of affordable housing. Thaler suggests an ordinance requiring developers to allocate a percentage of housing units to be affordable in order to receive any city funding. Community members can advocate at a local level for such ordinances, seeking to increase access to affordable housing in the midst of a luxury apartment boom.

Changes in city and state eviction processes can allow a more ethical approach to eviction. In Georgia, the eviction process can be completed in three weeks, allowing little time for appeal. Tenants also have no right to legal counsel during an eviction in most states. This creates a significant power imbalance in which landlords have representation and low-income tenants are left on their own. In

August of 2017, New York City passed the first law of its kind in the nation guaranteeing legal representation to any resident facing eviction with an income of 200 percent or less of the federal poverty level.[7] Similar efforts are launching in other cities. The Housing Right to Counsel Project in Washington, DC, began in 2013 as a collaboration between the DC Bar Pro Bono Center and the DC Access to Justice Commission.[8] The project provides free legal representation to subsidized housing tenants who are sued for eviction. In 2017, the DC City Council set aside $4.5 million to provide low-income tenants with a lawyer for eviction proceedings.[9]

At a federal level, citizens can advocate for increased federal subsidies for affordable housing. As mentioned in part one, every state in the United States is experiencing a severe shortage of affordable housing. The stress of deplorable living conditions, fear of eviction, and chronic homelessness is devastating to health. For many, like Darrell, a small, safe place to live is the difference between sickness and health. Creating thriving neighborhoods starts with people having a place to call home.

However, the impact of housing and environment on health isn't determined by housing alone but also the environment where that housing resides. Tre, who shared her experience of living in the projects, has thought a lot about the importance of place. She is a "Grady Baby," signifying her birth in the iconic downtown Atlanta Grady Memorial hospital, and has remained in Atlanta her entire life. She has lived in a variety of neighborhoods in and around Atlanta, including the now demolished Techwood Housing Projects. When asked what makes a neighborhood healthy, she pauses and says, "A place where people are kind to each other, . . . where neighbors talk to each other." Her answer struck us in that it had nothing to do with the affluence of the neighborhood. Her experience in neighborhoods throughout Atlanta was largely shaped by social cohesion and a sense of community. Neighborhoods can be transformed without poorer people moving out and richer people moving in. Creating healthy neighborhoods begins with a built environment that fosters community.

Michael Halicki, executive director of Atlanta nonprofit Park Pride, finds safe parks and greenspace to be one such strategy.[10] "The question is," he explains, "do people feel safe being outside in their neighborhood? If there is either a real or perceived fear of crime, people are more likely to stay indoors and be more sedentary. Further, if their housing conditions are substandard, they may be chronically exposed to respiratory issues from mold. The flip side is that a safe park can be a catalyst for health in a neighborhood. In addition to providing spaces for physical activity, parks promote social cohesion and community ties. An active, well-maintained park is a signpost of health in a neighborhood."

According to a 2017 article by the National Park Service and Centers for Disease Control, the public health benefits of parks and trails are broad and crosscutting.[11] "For individuals, benefits include providing places for physical activity, improving mental health, reducing stress, providing connections to nature, and increasing social interactions. Parks and trails can simultaneously provide venues for community events, activities, and public health programs and improve the environment."[12] An effective step for community-development efforts should include creating or cleaning up a local park. Beyond the benefits of the park itself, it would help bring attention to other important built-environment features that are in need of improvement.

Examples of other built-environment features that play a role in community health are things like well-paved roads and sidewalks, quality housing stock, well-maintained traffic lights and signposts, well-lit streets and gas stations, quality grocery stores, good schools, and proximity of police and fire stations, and so on. It's not only the presence of healthy features but also the absence of neighborhood toxins such as waste facilities, landfills, or predatory businesses.

These types of negative assets often get placed near poorer communities because the residents are less equipped to organize against community-planning decisions happening at higher levels of government. In fact, a recent study by the University of Michigan and the University of Montana researchers has shown that "hazardous waste sites, polluting industrial facilities and other locally unwanted land uses are disproportionately located in nonwhite and poor communities."[13]

Most of us share similar desires for our neighborhood regardless of socioeconomic status. We want affordable housing, convenience, good schools, access to public transportation, low crime rates, and neighborhood amenities. These factors are also determinants of health. The Robert Wood Johnson Foundation's action steps to achieve health equity emphasize the importance of environment, including environmental improvements to reduce health inequities, improving community capacity to address obstacles to health equity, and focusing efforts on those with the worst health and fewer resources.[14]

Pursuing health equity is not only about improving the lives of those in need. From a bird's-eye view of a whole city, or metro area, we are all affected by the health of depressed neighborhoods. According to Mindy Fullilove in her book *Urban Alchemy*, cities (like the human body) are a system.[15] If one part is fractured, the whole city is fractured, causing blockages. It's as if blood can't flow in the city. She insists that in order to repair the fractured state of urban cities, the whole city has to get involved with a common mission and strengthen the region. This is no longer just a job for those in the business of urban renewal, it should matter to everyone.

Michael Halicki explains that a lot of people who get frustrated with their local environment do the logical thing and move out. Yet a few committed advocates stick around, dig their heels in, and give everyone who will listen an earful about why their park, and their neighborhood, deserve better. Halicki says you don't need an army to change a park, just a handful of advocates who believe in the power of greenspace and will carry that message forward until they get results. However, they can't do it alone. It takes citywide commitment and political will to reengineer built environments. What is the motivation for such a sea change? The fact that better cities make life better for everyone. Getting cities right can yield major dividends in the form of efficiency, reducing greenhouse gas emissions, generating savings, and encouraging new business growth and development.[16] In other words, well-built neighborhoods and cities make economic and financial sense.

Cole Thaler uses an illustration of housing needs that mimics Maslow's hierarchy of needs. People first need basic shelter before they can be concerned with utilities. Electricity and running water are needed before appliances. People don't reach the tip of home decor if they cannot afford furniture. Society operates in much the same way. In order to create healthy communities, we start by housing everyone. We commit to viewing basic shelter as a human right and allocate the resources necessary to make sure our neighbors are not sleeping on the street. Next, we create policies and legal support systems to ensure that people are living in safe environments free of toxic mold, contaminated water, and insect infestation. Then we can begin to consider the makeup of individual communities and cities. We start with the basics such as food access, emergency services, and hospitals and health care. Then we design a built environment to facilitate health, including green spaces, recreation, educational programs, and community events. Solutions and strategies exist at each level of this hierarchy. If we start at the base and move upward, we can create a society where everyone has a safe and healthy place to call home.

One evening in early spring, we went to visit a patient in the hospital. Michael was a long-established patient at Good Sam, and we had become invested in not only his care but his life outside of the clinic.

Michael loves giving hugs and regularly updates us on his friends who need help and what he would like us to do when they come to Good Sam. He brings us cards for holidays and writes down everyone's name in his notebook so he will remember. He genuinely loves people and thrives on social interaction. Michael is also homeless and sleeps outside. The winter preceding this spring evening had been particularly cold. When Michael came to the clinic, his clothes were usually damp and his skin was dry and cracking from exposure. Over the winter months, Michael's cheerful disposition was replaced with anger and aggression, which eventually resulted in hospitalization.

On the evening of our visit, Michael met us in the visitor lounge with hugs and a huge smile. He was dynamic, telling us about everyone he had met, how great the food had been, how much he had been reading his Bible, and his plans after discharge. His brightness and enthusiasm were both heartbreaking and encouraging. While we were thrilled to see him this happy, we were also struck by how quickly a few nights inside had changed his demeanor. A few nights of sleeping in a somewhat sterile and impersonal environment with meals of hospital food had been transformative. Michael was free from the stress of wondering when and how he would eat again. Warm, dry bedding replaced cold, damp surfaces, and quietness replaced the noise of night in Atlanta. At that moment, these felt like such obvious solutions. These were simply basic necessities without which any of us would become stressed and irritable. It was heartbreaking to think that this was the difference between wellness and a downward spiral for Michael.

Yet we left the hospital hopeful that a healthy, vibrant Michael was entirely possible, and we were more determined to find the resources to make that possible. We left with a reminder that addressing social determinants of health can be restorative, and that it isn't too late for Michael, or the many like him, for whom a healthy life is possible.

CHAPTER 17

REENVISIONING HEALTH CARE

If only everybody could realize this! But it cannot be explained.
There is no way of telling people that they are all
walking around shining like the sun.

THOMAS MERTON

One Friday in August 2015, we opened the clinic doors with a plan to try a new model of health care delivery. A few months prior, a conversation had started about Atlanta's homeless and how our clinic, while serving almost entirely uninsured patients, saw relatively few individuals experiencing homelessness. We started with a question: If we want to provide care for people experiencing homelessness, what do we need to do differently? This day in August was our first attempt at answering that question.

Despite an increased national focus on patient-centered care, health care systems often operate in ways that best serve the health care systems themselves and the people employed by them. In outpatient care, we schedule appointments at specific times and schedule separate appointments on different days when specialty consultation, behavioral health, or dietician visits are needed. Rigid schedules and multiple appointments spread out over several days do not promote health care access among homeless populations. The most significant barrier to care for the homeless is cost.[1] Even a reduced-cost medical visit is not affordable for someone without any source of income. A medication on the four-dollar list at a retail pharmacy might as well be $200. The National Needs Assessment found that lack of transportation, lack of knowledge about available services, and lack of insurance were also barriers to accessing care.[2] Sustainable health care solutions for homeless populations require addressing all, not just some of these barriers, presenting a major challenge to any health care system.

Homelessness is elusive to define and quantify. The Department of Housing and Urban Development (HUD) establishes four categories of homelessness.[3] These categories include individuals and families with a primary nighttime residence that is not designated as "a regular sleeping accommodation for human beings"; those sleeping in shelters or transitional housing; those in institutions (such as hospitals) who resided in shelters or places not meant for sleeping prior to admission; and those fleeing domestic violence who lack support systems to obtain other permanent housing. On a single night in 2016, 549,928 people in the United States were experiencing homelessness.[4] Of these 68 percent were staying in shelters and 32 percent were in unsheltered locations. This number doesn't include homeless individuals who are doubled up, meaning temporarily sleeping on the couches or floors of friends' homes. In Atlanta, just over four thousand people were homeless in January 2016, and 21 percent of them were unsheltered.[5]

Individuals experiencing homelessness experience disproportional health risks. Poor health can be both a cause and a product of homelessness. Among homeless adults, 28 percent have a severe mental illness and 22 percent are recognized as physically disabled.[6] Without adequate social support and health care, physical and mental illnesses can make it impossible for individuals to sustain employment and, as discussed in part one, even disability payments are often not enough to secure housing. Homeless individuals are three to four times more likely to die prematurely and have longer stays when hospitalized.[7] They also experience higher rates of communicable diseases such has HIV, hepatitis, and tuberculosis.[8] Perhaps most alarming, life expectancy in the homeless population is between forty-two and fifty-two years, twenty to thirty years less than the general population.[9]

With this in mind, in August we piloted a clinic dedicated to providing care to homeless individuals. We approached some community partners, including a downtown ministry, a recovery program, and a local shelter, about the needs of their communities and our hopes of partnering to offer health care. The monthly clinic would offer primary health care and behavioral health. Six patients arrived for that first clinic. One cheerful man with a huge toothless smile shared with us his story of fleeing to the United States with his parents as a young child during World War II. His childhood morphed into a tortuous adolescence, and he underwent experimental psychiatric treatments. On his own, without support or treatment, a series of bad decisions landed him in prison, where he spent most of his adult life. Freed due to the persistence of activists, he told us his favorite activity is now whistling to and feeding the birds. Fifteen minutes into the office visit, we still hadn't asked him a single question about his health. Another patient left yelling at our psychiatrist, her voice echoing down the hall. We decided the clinic was meeting some unmet health care needs in the community, and we planned for the following month.

The program grew into a weekly clinic with a team of primary-care providers, dentists, counselors, a certified peer specialist, and psychiatrists providing care to about forty people each week. Now, preparations begin before the sun comes up. Our van pulls out and heads to a park downtown to pick up folks ready for care. The Good Sam team sets up a breakfast buffet in the lobby, and volunteers assemble care packages for every patient. A local stylist transforms an exam room into a salon for free haircuts. When the van returns to the clinic, patients fill up plates with breakfast food and sign in. Using an integrated primary-care delivery model, patients begin their office visit with a primary-care provider. As needed, patients can see a psychiatrist or counselor, allowing the providers to collaborate and create a plan of care together. Patients also have access to dental care. An hour into the morning a social worker from a partnering organization arrives to conduct housing assessments. Finally, patients leave with bottles of medication filled at the dispensary.

Most Fridays the operation looks like well-organized chaos. One of our psychiatrists fills out a HUD McKinney form to verify a patient's need for supportive housing. One primary-care provider performs a patient's first complete physical exam in ten years while another makes a plan for a diabetic patient with a severe skin infection who won't set foot in an emergency room. At the end of the hall, our counselor leads a guided-meditation session with a few patients who are waiting for an appointment. Another primary-care provider strategizes with the dentist about how to manage a patient's blood pressure when he needs all of his remaining teeth extracted. The medical assistants bandage wounds, draw labs, and search the storage room for donated coats. The farm team leaves the field for the day to ride the fifteen-passenger van as it picks up patients from the park and downtown shelter. The development team focuses on stocking the breakfast buffet instead of calling donors for the morning.

We have come a long way from that August morning in 2015, but the process has required us to relearn everything we thought we knew about running a health center. Providers show up without a schedule, willing to see whoever shows up for any number of concerns. The front desk staff has learned to triage and schedule a mob of patients rushing in at the same moment. Providers sometimes spend more time talking about housing, recovery meetings, and public transportation than medications. Quality assurance checklists are replaced by a simple question: What do you need to feel better? Dr. Bill Warren likes to remark that all this activity reminds him of the old TV show *Cheers*: "It's where everybody knows your name and you belong."

In some ways, our naivety and vulnerability have promoted the success of the program. We didn't launch the homeless clinic with a particularly well-designed plan or extensive expertise in homeless care. Rather, we started with a commitment to deliver compassionate care and a willingness to change to meet the needs of those who came. For Janet, a morning at Good Sam included visits with primary care, counseling, and psychiatry. "You know, if people like me have to be here all day, you could at least give us some food," she said. We started bringing bagels and juice to the next clinic. Soon more staff members and volunteers started bringing food and soliciting donations from local grocery stores. Now patients fill their plates from a complete buffet every week. Our providers take a similar approach. People most often know what they need to get better. Jerome had been in and out of the emergency room for diarrhea, unexplained weight loss, and insomnia. Each time he was given an HIV test and then told to follow-up with psychiatry to manage his bipolar disorder. "I know I'm bipolar," he told us, "but something else is going on." He was right. Now, six months into the treatment for a thyroid disorder, he is sleeping through the night and has gained fifty pounds.

Good Sam's homeless clinic is only possible due to the commitment of volunteers, donors, staff, and community partners. On any given Friday, nearly half of our medical team is made up of volunteers. From our rotating volunteer providers to the students who volunteer as medical assistants, to our volunteer psychiatrist and counselor, our growth has been possible through a growing volunteer base. Having a team of committed volunteers who rotate weeks nearly doubles our capacity. We recruit volunteers through our website (which has an entire page dedicated to volunteer opportunities), a competitive internship program for students planning careers in medicine, and word-of-mouth.

Despite volunteer support, there are also ongoing expenses to running the program, such as the Good Sam staff salaries, medications, and lab costs. The Good Sam development team collects patient stories, offers tours on Fridays, and matches donor interest to specific needs, creating a funding stream specifically dedicated to this program. This starts with our employee giving campaign. One hundred percent of Good Sam employees give to the organization, and all employee donations go directly to support the homeless clinic. Next, individual donors support specific needs. One family donated the fifteen-passenger bus, allowing us to transport patients to and from appointments. Another donated a year's worth of supplies for the care packages. The development team also submits grants specifically earmarked for our Friday program.

Our Friday clinic would not be possible without the commitment of our staff. The administrative team leaves their farming, accounting, fundraising, and operational tasks for the day and instead sets up breakfast, rides the bus, and hands out care packages. Another team member sits in the center of the nurses' station keeping track of who is being seen and what additional consultations each patient needs that day. This information is also entered into a massive database that can be used to generate data for grant applications and reporting. The dispensary techs count out pills and label bottles for providers to dispense to their patients. They also keep track of which medications are running low so they can be ordered each week. The medical staff and providers surrender their regular schedule and see patients as they arrive. Rather than a tightly planned day, they adapt moment to moment, meeting the needs of each individual.

Collaboration with partnering organizations allows us to create a program that moves beyond health care and addresses social determinants of health. While interdisciplinary collaboration within the health care team has become the standard of practice, if social determinants are driving health outcomes, then collaboration must go beyond the health care team. Educators, social workers, civic leaders, pastors, and community organizers are equally important in creating healthier neighborhoods and eliminating life-expectancy gaps. This is vividly illustrated in our Friday program. Each week, we send out emails to local case workers, a deacon, a pastor, and the organizer of Atlanta's largest recovery meeting in the city specifically designed for people experiencing homelessness. In these emails we list individuals who have been referred by their prospective organizations who are due for a follow-up.

As these community members and case workers interact with people during the week, they will remind them of their need to visit Good Sam. Occasionally, they email us when they notice changes in one of our patients. Their observations have alerted us to symptoms we would have completely missed in a twenty-minute office visit. These community partners have helped drive patients to imaging and specialty appointments, and they store their medicine for them to prevent it from getting lost. The trust and community within each of these organizations are healing people in a way medication never could. Two community partners also come to Good Sam on rotating Fridays to conduct housing intake assessments and provide case-work services. As a result, several of our patients have found permanent supportive housing. These partnerships are the foundation of our program and offer a vision of the success clinical programs can have when collaborating with nonclinical interventions and support systems.

Friday clinic also illustrates that creating healthier communities is not only the work of people in the medical field. One particularly powerful example of the significant impact of nonmedically trained healers is the Certified Peer Specialist (CPS) program. Certified Peer Specialists are individuals who have personally experienced mental-health illness and partner with peers to offer hope, role modeling, and assistance in their personal recovery journeys.[10] The project started in the early 1990s in Georgia by individuals living with mental-health illness who were dissatisfied with the care they were receiving. Ten years later the group held their first two-week class, preparing individuals to become CPSs with the vision that wellness and recovery are possible. The increased utilization of peer specialists is also a solution to national challenges in mental-health care. Peer specialists can increase conversations about mental-health illness, decrease stigma, and offer support in the recovery process.

Sam Rapier, CPS training coordinator, explains that his lived experience has made him passionate about recovery.[11] He was diagnosed with schizophrenia triggered by his mother's death shortly after finishing his second master's degree. When he started hearing voices he sought health care, but medication didn't stop the voices and he often found the side effects worse than the voices. He recalls a turning point when standing in a Kroger grocery store one night, hearing voices coming out of the speaker. "I didn't run out of the building," he remembers. "I realized I was able to make a change." The voices never left, but he has learned to live with them, shutting them out and identifying triggers. He has dedicated his life to helping people do the same, "realizing the changes they can make and affirming those changes."

Mariam Abdul-Aziz, CPS certification coordinator, nods in understanding as Sam tells his story.[12] Her own lived experience involves years of mood swings and severe depression. Her first interaction with the medical system came while she was working in Africa. One of the doctors on her team suggested that she might be dealing with a mood disorder. "Will I be okay?" Mariam asked. "Oh, you'll be fine," the doctor responded, "with the right medication and the right support." This interaction was profound for Mariam because the doctor's response offered hope. "This experience [with the medical system] is very rare," Mariam tells me. Too often, the medical community's response to mental illness is anything but hopeful. Sam explains that hearing voices is immediately assumed to be a problem, but no one ever inquires about the content of those voices. Mariam describes that a diagnosis of a mental illness is often accompanied by a list of things that are supposedly incompatible with severe mental illness, such as gainful employment, a successful marriage, or raising a family. Through the CPS training, individuals like Sam and Mariam learn how to use their personal stories to share a message of hope.

Sam and Mariam describe sharing stories as a transformative process. There is a mutual understanding born from a shared lived experience. "It's almost like speaking a language," Sam explains. "You kind of get it because it's the same language. It's not that people have walked down the same path—they don't have to. It's about feeling the feelings." Certified Peer Specialists help people find healing through their stories. Their authenticity creates a space for people to feel understood and validated. The program also helps individuals identify their strengths. "We ask, 'What's *strong* with you,' not 'What's *wrong* with you,'" Sam explains. Mariam describes how her mood disorder has brought wisdom to her life. The experience "made me a much more compassionate person," she tells me. "I wouldn't change anything for the world."

Sam recounts discussing some of his thoughts about the voices with his husband. "I can't go where you are," his husband told him. "There is nothing in my experience that can help me deal with this." His sentiment is shared by many health care providers treating mental-health illnesses. There is no amount of reading or studying or degrees that can enable providers to connect with people the way Sam and Mariam do through their lived experiences. Sometimes people benefit from medication, counseling, or psychotherapy during their recovery. However, people always benefit from support, community, and understanding. Finding partners beyond the bounds of the traditional medical system has created new opportunities for healing.

When we added a CPS to our team on Friday, we immediately saw the power of sharing recovery experiences. One morning, a man in his early twenties was referred from a local shelter by his case worker. He told his provider about his racing thoughts, inability to concentrate, and risks for sexually transmitted infections. However, when our CPS sat down with him the young man started to share his dreams for the future and his frustrations about falling short of his personal goals. Our clinicians can help him by screening for infections and prescribing medication to manage his racing thoughts, but our CPS is providing a safe space for him to dream and hope that the better life he envisions is possible.

While Good Sam's Friday program is dedicated to serving individuals experiencing homelessness, the rest of the week we provide care to individuals and families at risk for going with unmet health care needs. As a result, we are constantly adapting our processes and programs to increase their access to primary care. For example, we are open every Saturday so our many working patients can make appointments without losing a day of work. The mobile mammography unit comes four times a month to Good Sam so patients for whom transportation is a barrier to receiving their mammogram can receive screening before exiting our parking lot. We have learned that increasing access to primary care is about more than making it affordable.

Nonprofit, charitable clinics like Good Sam exist throughout the United States, forming a safety net for the 27.6 million uninsured.[13] Some are Free and Charitable Clinics, seeing primarily the uninsured with little to no government funding. Over twelve hundred clinics like these provide care for five million Americans each year.[14] Federally Qualified Health Centers (FQHC) are community-based health centers receiving federal funding and increased Medicaid reimbursement to provide primary care in underserved areas.[15] Over ten thousand of these centers provide care to twenty-three million Americans with Medicaid, Medicare, or no insurance.[16] Charitable clinics range from small volunteer-based programs operating a few nights a week to large centers with multiple locations and hundreds of staff. They are located throughout the United States, from rural Appalachia to downtown Chicago. Regardless of their differences, they supply critical primary-care services to people at risk for undiagnosed cancer, untreated chronic disease, and poor health outcomes. They also are dependent on philanthropic funding and volunteers. Even FQHCs can meet only part of their budget through government funding. The demand for these types of clinics is increasing while funding has dwindled. Consider being the volunteer your local clinic needs or donating regularly to their operating funds. Every donation makes it possible for more patients to receive care. Finally, advocate for funding at the state and national level. Let your representatives know that you want funding to be available to the clinics serving the most vulnerable in our nation.

While charitable clinics and FQHCs provide essential, high-quality services, like any safety net, even when well-built and operating correctly, holes exist. Charitable clinics do not have the resources to treat cancer or stabilize a patient during a heart attack. In addition, as funding is cut, clinics have less money to put into educational programs, community outreach, and partnerships, limiting their ability to address social determinants. At Good Sam, significant fundraising effort and generosity is needed to provide services to a small portion of Atlanta's homeless population each Friday. While we are working to create a model of care that can be replicated by other clinics, without national reform, this is not a sustainable national solution.

Strategies for improving access to primary care, meeting national goals for screening exams, and achieving mental-health parity are needed yet inherently limited in our current approach to health care. While the quality of acute care in the United States is high, management of chronic disease in primary care lags in quality compared to other nations.[17] We spend more on health care than any other Organisation for Economic Co-operation and Development (OECD) nation, yet we are the only nation to leave citizens without insurance. In the United States the answer to the question, Does everyone have a right to health care? is still no. Health care resources are rationed by ability to pay rather than need. We leave millions of people uninsured and millions more without access to affordable care. We have implicitly accepted the fact that the poor and uninsured have shorter and unhealthier lives as a necessary consequence of our health care system. But it doesn't have to be that way. Providing health insurance and access to primary care looks different across the OECD nations. From national health care in the United Kingdom to a single-payer system in Canada to private sickness funds with government regulation in Germany, each country has found a way to guarantee health care to all its citizens. The United States can do likewise. However, it will require a strong, collective voice of citizens demanding that health care be treated as a right versus a good available only for those who can pay. Health equity is possible only when we decide our neighbors have a right to the same level of health care as we do.

Increasing the number of people with health care and supporting the growth of primary care through improved reimbursement strategies will be transformative for charitable clinics. We dream of a day when providers can spend less time printing coupons and completing prescription-assistance applications to help patients afford their medications, a day when providers don't have to call multiple offices and help patients complete financial applications for basic procedures and specialty consults, a day when the funding raised by the development team isn't needed to cover basic health care services. We could spend more time talking to our patients about utilizing their community park, locating healthy housing, buying and preparing produce, and health screenings. We could spend more time developing strategies to connect patients with community services, create health-education programing, and conduct community outreach. If our patients had health insurance, we could focus on addressing the underlying factors really making our patients sick.

CHAPTER 18

RX FOR CHANGE

AN APPROACH TO ACTIVISM

Nothing about us without us.

ACTIVIST SLOGAN

NATIONAL APPROACH

On the staff bulletin board at Good Sam hangs a cartoon strip carefully torn from a newspaper and yellowing with age. In the first frame the character gets out of bed and reaches for his to-do list. The to-do list says, "Save the world." The character pauses and then thinks, "Good thing I got up early." When we think about social determinants of health and life-expectancy gaps, the solution seems like saving the world: elusive and overwhelming. However, our work in Atlanta has convinced us that change, while difficult, is very much possible.

This book has focused on local interventions as strategies for addressing social determinants to achieve health equity. This is, in part, because our personal experience and expertise lie in this area. Second, local initiatives are often the most effective at creating community change and can be incubating grounds for larger social movements. Housing First, for example, started as a pilot program in New York and is now a national strategy for eliminating homelessness.

The social justice focus and equity-review process employed in King County are being explored in other US counties with national implications for how we make policy decisions. However, we would be remiss not to emphasize that achieving health equity and narrowing the life-expectancy gap is not possible without national-level reform.

In their discussion of social determinants of health and the built environment, Amy Schultz and Mary Northridge note that economic and political orders, legal codes, human rights doctrine, and social ideologies in the United States have created concentrations of poverty and wealth, often leaving racial and ethnic minority groups with less access to social, political, and material resources.[1] The implications of this are substantial. While well-designed local interventions and community-level reforms can mitigate the damage, we cannot change the trajectory of US health inequity without addressing national policy. For example, housing policies need to increase the availability of affordable housing and fund housing vouchers that allow for independent living rather than homeless shelters. Living-wage laws promote gainful employment and allow working adults to support their families.

Fortunately, social ideology in the United States is slowly starting to change. Sixty-five percent of Americans believe that "one of the big problems in this country is that we don't give everyone an equal chance in life," up from 53 percent in 2010.[2] The majority of Americans favor a more equal distribution of money, housing subsidies, and raising the federal minimum wage.[3] The next step is for these social ideologies to inform our policy making. This requires widespread participation in advocacy and increasing participation among diverse populations.

Engaging in advocacy is not easy. Whether you are balancing a career, school, family, or simply all the management daily life requires, finding time to engage in the political process is difficult. However, you don't have to be an expert to make a meaningful impact in national policy. Start by finding a topic you care about and take the time to become knowledgeable in that area. For Breanna, the obvious fit is health care. When health care legislation is up for a vote at the federal level, Breanna makes calls and send emails, sharing them with friends and colleagues so that they can send similar ones. Veronica is particularly engaged in the Georgia Charitable Care Network, promoting the success of clinics like Good Sam. However, we recognize that economic, housing, environmental, and educational policies are even more important in promoting health equity than health care policy. Yet we are not experts in all of these areas. Clearly, there are numerous social determinants needing champions for change, but no one can do everything.

Rather than becoming overwhelmed by issues in which you lack experience or expertise, find an organization or individual who is passionate about the topic and aligns with your values. Sign up to receive policy briefs from an organization you trust. When they send an email asking you to make a call, use their script and call. Keep in mind that you can advocate for policies that promote health equity without aligning with a particular political party. National, bipartisan commitment to health equity is necessary for its achievement.

Progress is also advanced when diversity in leadership and advocacy increases. Poor people and members of minority groups are less politically active than those with social and economic advantage.[4] They are also less likely to mount opposition when policies are not in their favor and services are reduced in their area, creating a situation in which elected officials are less incentivized to respond. People who are experiencing homelessness, moving from job to job, or struggling to make it to the end of the month with enough food are less likely to have the time and capacity for advocacy. Matias Valenzuela, the director of King County's Office of Equity and Social Justice, emphasizes that building community capacity is a core component of their strategic plan. "The question with any initiative is:

How are you building community capacity?" he explains. To accomplish this, the county government partners with community organizations to fund capacity-building activities. The goal is not simply to improve an at-risk neighborhood but to give residents the power and resources they need to advocate for themselves. In King County this commitment also resulted in an internal evaluation of local government. The county examined its hiring practices to promote more diversity in all levels of county government and particularly in leadership positions.

We think about what capacity building might look like at Good Sam. Providers talk to their patients about healthy living choices, housing, and employment, but how can we equip them to advocate for policy change? Patients have asked us if we think health insurance is worth the money. We were regularly asked if the Affordable Care Act would make any difference. These questions continue to challenge us. We would never withhold a medication or treatment that could benefit patients, yet we have not done much to help them understand how many of our nation's health care policies work against them.

Just like social ideology, political and economic systems can change. The same systems that have sustained inequality in the United States have the power to eliminate inequity and disparities.[5] Collectively, we have the ability to restructure our systems in ways that promote the health of all of our neighbors.

COMMUNITY APPROACH

We have seen this scenario play out so many times, and yet it is still surprising. Imagine sitting in a large conference room with about a hundred other people united around their common concern for three rough neighborhoods in West Atlanta. At 7:15 a.m., university presidents, fortune 500 CEOs, city council members, ministry leaders, nonprofit CEOs, and more had gathered to learn how they could help. A small section of the community was about to receive $400 million in funds to improve housing conditions, educational offerings, and social services. We expected to see a lot of joy and gratefulness and unity.

But right off the bat the morning began with a devotion by one of the local residents who reminds everyone that he has lived in the community, along with others in the room, for a long time. He says the community has been working hard for years, and their work should not be overlooked. Even though we both lived in rough neighborhoods at the time, for many of the same reasons as him, we suddenly felt ashamed that we didn't live in *his* neighborhood. We looked around wondering if any other "nonresidents" felt uninvited.

Then the floor opened for people to speak what was on their hearts before the business portion. Now a long-term, multigeneration community resident stood and reminded everyone that they should listen to "the people," who had been ignored for a long time. We then felt both unwelcome and guilty.

Finally, the business portion started and a thorough presentation was made by a local organization to celebrate a huge grant that had been awarded and outline the plans for how it was to be spent. It was thoughtful and strategic, with an eye toward long-term sustainability and empowerment of the neighborhood. But as soon as the presentation ended, the shaming began. Residents took turns complaining that the community had not been involved in the plans, one resident specifically taking issue with a safety initiative. As the speaker's face turned red and her shoulders slumped in defeat, we wanted to stand up, blow a loud whistle, and play referee. We have sat on *both* sides of this dysfunction and misunderstanding countless times.

We wanted to explain to the speaker why the community was mad. While the nonprofit had organized "community meetings" to gather resident input, they had not regularly attended the community's meetings. This may seem like semantics, but it's critically important.

The community had a neighborhood association and a local neighborhood planning unit, and this is where they did their business. If you didn't come to those meetings, you weren't truly inviting community feedback. The expectation of outsiders is that they attend local meetings for a long time and then seek the permission of the neighborhood for any change initiatives. The neighborhood does not want to be on outsider's agendas at outsider's meetings. They want the reverse.

We wanted to tell the community residents to consider the benefit of having one of the roughest, most neglected streets in their community have five vacant properties replaced with five new affordable homes. We certainly understood the skepticism toward outsider involvement, but as residents of similar communities it seemed crazy that this was even a point of contention. We wanted to say gently that while they had good intentions and right desires, they were now standing in their own way. An entire room of Atlanta's movers and shakers would walk away scratching their heads and wondering why the neighborhood was angry about getting better.

To local residents active in the redevelopment of your community, consider taking a more pragmatic and less personal approach. If you shame and fight those who are willing to invest resources in your neighborhood, you do so at your own peril. Seek ways to communicate your vision for your community in ways that invite partnership and collaboration. You can intentionally build relationships with people who have influence, resources, and personal interest in your neighborhood and negotiate with them. Go to them, articulate what you want and why you want it, give them the benefit of the doubt, and when they do something right—praise them for it. Realize that all forward progress is a mix of good and selfish intentions, and find a way to maximize the good on behalf of your neighborhood.

To funders and civic leaders seeking to support community change, please remember that you do not know what a neighborhood needs better than the residents themselves. Because you start with the upper hand of having influence, capital, federal support, and the like, you must double down on your efforts to defer to the community and follow their local processes for change. You must move slowly, take time to build trust, and respect the community leaders (regardless of whether their education, knowledge, or capacity matches your own). Don't send your entry-level intern to the community meeting, instead send an executive leader. Don't offer gifts or plans with lots of strings attached and hoops to jump through. Consider how you can build community capacity and promote leadership from within the community. Most importantly, be transparent about the economic drivers pushing your project forward. Help residents preserve the history of their community while informing them of real economic changes on the horizon.

Throughout the process of community improvement or redevelopment, it is critical that community residents and outside funders embrace one another and learn to work together. Transforming neighborhoods is difficult and expensive work and requires partnership with lots of people and agencies to accomplish even small goals. Residents can often benefit from the resources and input of those who are connected, who come from different backgrounds, who have capital and vision, and who might even have different agendas. Funders can equally benefit from respecting the local community and linking arms to ensure their efforts are successful. Once this kind of cooperation begins, successful community development becomes possible.

INDIVIDUAL APPROACH

Countless times we have found ourselves moved by societal problems and challenged to make a difference yet at a loss for where to start. The larger the problem, the more insignificant our role as individuals appears. However, we are convinced that creating healthier communities starts when each of us as individuals understands the causes of health inequity and considers the health of our neighbors in the decisions we make. As we conducted interviews for this book, we asked each person the question, If you could tell people to do one thing to address education, homelessness, unemployment, and so on, what would it be? From their responses and our personal experience, we have created a list of ways individuals can commit to changing and laying the groundwork for healthier communities.

Listen. We often hear community activists described as being a voice for the voiceless, but this is not really accurate. It is not that oppressed and vulnerable communities don't have a voice but rather that people are not listening. How are you actively listening to people outside of your social circle and immediate community? Attend lectures, community events, and religious services that expose you to ideas outside of your worldview. Follow a blog from someone with life experience different from your own or who cares about issues you are less educated about. Intentionally seek opportunities to spend time with people of different socioeconomic statuses. Listen to their stories, and when confronted with your personal biases and prejudice, listen even harder. Most importantly, when you are trying to help, first listen to those you seek to help. People know what they need.

Diversify. Take steps to increase your interactions and relationships with people who are not just like you. Meet people whose race and culture, religious view, political viewpoint, educational status, and socioeconomic status are different from your own. Consider joining a community organization or book club where you can hear from people who do not look or think just like you. Seek friendships built on mutual respect and benefit. Next, foster diversity within the lives of those in your social network. Host a dinner party, parent group, Facebook group, or an event where people can meet and exchange ideas. One interviewee explained that as a black Muslim woman she has lots of diversity in her life but realized that none of those groups of friends overlapped. She held a party and intentionally invited everyone, overlapping the diverse groups in her life.

Tell stories. First listen, but then tell your story and create spaces where others can tell theirs. Use whatever platform you have to provide a public space where others can share their experiences, particularly welcoming those whose stories are most often not heard. Share your personal experiences with local leaders and legislators. Detailed data charts, well-designed studies, and thorough analysis are often unable to communicate what can be said in a single effective story. We learn through the stories we tell one another. Social media can be a powerful storytelling platform, but stories can easily change from vulnerable and convincing to ranting. Accompany your stories with examples of positive action, and maintain enough humility to listen to the voices who have experiences different from your own.

Invest. Locate the low-income neighborhoods in your area and invest in them. Find a local organization that has a meaningful impact in that community and donate to them. Shop in the community and support local businesses. When local decisions about transportation, school zones, or voting districts are being made, ask first about the impact of those changes on low-income neighborhoods and commit to advocating for decisions that strengthen those communities.

Volunteer. Find an organization with deep roots in the community you hope to impact and ask them what they need. Structure the way you volunteer around the needs of the organization, laying aside what might feel or look better for you. Commit to long-term support, recognizing that the needs of the organization may change with time. The faithful commitment of a long-term volunteer is a significant asset to a service-oriented or nonprofit organization.

Encourage. Seek to better the lives of people through encouragement. Model kindness in interactions and be present when given the opportunity to listen or walk alongside someone in life. Presence does not require a shared lived experience or even understanding. Yet when you are present with others you may find a commonality you would have otherwise missed. Particularly invest in young people. Find opportunities through mentorship. Support early childhood education and public education with your voice and your vote.

Work. Consider pursuing a career that enables you to contribute to healthier communities.[6] Our careers at Good Sam enable us to spend our working hours promoting health equity in our community. You might consider working in local government, like the King County Office of Equity and Social Justice, to affect policy change. By working for a foundation or university, like the Virginia Commonwealth University or the Kaiser Family Foundation, you can supply the data and reports needed by advocates and policy makers. You might run a youth-mentoring program or develop a thriving business that employs individuals leaving prison. Your skills, interests, and livelihood can promote health equity.

Engage. Change cannot happen without individuals engaging in policy decision-making. Engagement in policy making does not have to be political. Start by attending neighborhood association meetings. Follow your city council meetings and attend forums and town hall meetings as much as possible. Evaluate change through the lens of health equity: Will this decision make my neighbors, particularly those in low-income communities, healthier? Save the contact information for your city council members, district legislators, and state legislators in your phone and start with a goal to call once a month to advocate for something you care about. Finally, talk to your friends and family, post on social media, and encourage someone else to join you.

Step aside. Recognize your limitations within any service you are delivering or change you are trying to ignite. Sometimes you are not the best person to fix the problem, but you can create the support systems and opportunity for the people who are. Listen to those who are most affected by disparities, inequities, and systems of oppression, and then identify your role. Listen to what people need and what they don't. You are especially equipped to fix yourself and your immediate community first. If you are white, identify and call out racism within yourself, friends, and family. If you are a person of faith, challenge your congregation to be advocates for the oppressed and disadvantaged neighborhoods in your city.

Anthony, the community outreach manager at Good Sam, is a particularly good example of the powerful impact one individual can make on the health of their community. Anthony lives in Good Sam's neighborhood with his family, and they live a very healthy lifestyle, eating a vegetarian diet and exercising regularly. Anthony is involved civically in all the local neighborhood organizations and schools, and has personal friends who either serve in city council leadership or are closely connected. When his neighborhood needs something, he advocates for change. He has also established a youth development and mentoring organization, the Logan Wilkes Foundation, named after his son, which gives kids in the community a vision for positive futures. Through the foundation he also teaches work ethic and personal responsibility to the kids by having them regularly staff the concession stands at local football and baseball games to raise money for the foundation's programs. Through his position at Good Sam, Anthony is able to run health education and community outreach programs tailored to meet his neighbors' needs.

Above all, Anthony is an extremely hard working and charismatic person who maintains a positive attitude yet acknowledges there is much that is still broken. He works with a passion and urgency that is contagious. Donning one of his favorite green T-shirts that simply says *Plant, Grow, Eat, Repeat*, he shows up to work each morning ready to take the next step to make health equity a reality in his community.

Put a few Anthonys together and the famous Margaret Mead quote will come to life: "Never doubt that a small group of thoughtful, committed citizens can change the world. Indeed, it is the only thing that ever has."

EPILOGUE

TOWARD A BETTER WAY

BREANNA

The cartoon on Good Sam's bulletin board reminds me of the idealistic, naive new graduate I used to be. I remember thinking that if I just got up earlier, worked harder, or called enough elected officials I could save the world. But the world already has a Savior, and it isn't me. A few degrees and a prescription pad aren't remedies for the problems plaguing our nation. Poverty, racism, segregation, homelessness, unemployment, and other systems of oppression and injustice are making our neighborhoods sick. Nothing short of a national commitment to change and an army of advocates to lead it will create a nation in which all people have a chance to be healthy.

I still struggle with my personal role in dismantling systems that allow such grave health disparities to exist. After Good Sam's homeless clinic on Friday, I get into my relatively new minivan, pick up my daughter from a high-quality early childhood learning center, and drive to a home we own stocked with food and other necessities. I recognize that the social constructs and political systems that have paved the way for my privileged lifestyle are the same that have mounted devastating disadvantages for my patients. I am not sure how to live with this reality, but I've resolved to let it continue to make me uncomfortable.

I'm also not the same frustrated nurse practitioner a year into my career questioning whether I can make any difference at all. While my work daily reminds me of the brokenness of this world, I also see examples of human resiliency and a remarkable amount of good. I have been inspired by providers and community partners who are making a lasting impact in their communities. Henry and James are both working steady jobs and living in apartments of their own. Kelly works as an accountant, and Barbara received supportive housing and moved to another community. A patient this week informed me that he is moving out of a shelter and into his own place. He's been offered a position with full benefits as a case manager and peer counselor. His lived experience and willingness to share it will become part of someone else's recovery journey. Change is possible when opportunities for support and recovery exist.

However, not every story has a resolution. Janet and Michael are still sleeping on the street. Janet recently told me she was considering moving inside, and I offered to connect her with a case worker I trust. For the first time in our years of working together, she said yes. Michael has a great team of advocates, and I am looking forward to the day when I get to tell him we have a place for him to call home.

We are a long way off from health equity in the United States, but it is possible. The systems making my patients sick won't be cured in their lifetime or mine. When I sit down with a patient who is one hundred pounds overweight, I talk about how ten or twenty pounds can make a significant impact on blood pressure and blood sugar. Quality of life and health outcomes start improving with small changes. The same can be said of our neighborhoods. Small changes—a new park, the implementation of an equity-review process in local government, a grocery store in the middle of a food desert—can make meaningful differences in the health of a community.

Small changes can gain national momentum when increasing numbers of people demand change and commit to being part of that process. Be a part of that change process. Donate, make a call, diversify your life, or volunteer. Attend neighborhood meetings, vote in local elections, join the board of directors for a local nonprofit. Make a shift in your career, your commitment to financial giving, or your ideologies. The inequities in health and life expectancy created by social determinants of health are making our neighborhoods sick, and we must do better. Life-expectancy gaps close slowly, but they can close. We can commit to making our neighborhoods healthy, and we can let this drive our ideologies, policies, and interventions. And while I still struggle to find my place in this and grow ever more aware of my own limitations, of this I am sure: There is a better way for our nation to be healthy.

VERONICA

If I am honest, I am still trying to make sense of what happened (and why) in Southwest Atlanta. There are threads of clarity and purpose, most of which I have shared in this book, but just as many unanswered questions and lingering disappointments. I am healthier now, but still susceptible to triggers from the past, like the sound of fireworks (which remind me of gunfire) or ATVs revving down the street. I've come face to face with my own frailty, significant human suffering, and systems of injustice that make it hard to sleep at night. The world is a far darker place than I once believed.

Yet at the same time that my journey has taken me through a prolonged season of darkness, like a clear morning sunrise, the light is starting to break through—convincing me afresh of the power of Christ to raise what's dead inside of us. I find that despite my exposure to human suffering, I am driven by the lofty ideal of creating environments where health equity is possible so all people can be on a level playing field and have the opportunity to live a happy, flourishing, complete human life. Further, despite the great divisions that separate people from one another in our modern times, I am motivated to work toward the biblical concept of shalom (or peace) for our communities.

This renewed calling struck me most poignantly after my family moved out of Southwest Atlanta. Eric and I were still reeling from our experience and dealing with that grief on a daily basis. The hurt and strife that ensued made me wonder how we were going to make it. I found myself faced with a choice: to either respond as my natural self would respond based on how well my own needs were being met or repeat the mantra "love wins" daily and choose to be kind and serve others regardless of my own present satisfaction. I made it my mission to bring an aroma of love and shalom back into our home. The house became alive with my family's favorite things—super cheesy lasagna, homemade art, dark chocolate, encouraging words, hugs, and kisses. Over time this relentless effort, coupled with capable medical care, led to healing and restoration for all of us. To me, it felt like discovering a secret to life for the first time. In reality, it was simply learning about the grace and mercy provided through my faith in Jesus Christ. This is the same grace and mercy that Christ daily extends to me.

As I look back and think about the significance of my journey, I realize that in order to be a student of and advocate for health equity, I had to first know sickness. I needed to come to know it intimately—how it feels, what it looks like, and how it spreads. Without this firsthand bone-deep experience, my ideas or solutions for health equity would be shallow. Before Southwest Atlanta, I knew how to flourish in the suburbs, but those strategies would not have worked in a resource-drained environment for those in generational poverty. Now I have a broader view of the complex web of resources needed to heal urban neighborhoods. I am also encouraged by recent scholarship that shows human limits and suffering do not have to permanently impact one's prospects for thriving. In fact, with appropriate interventions, human frailty plays a critical role in the journey to living well. In the view of psychologists Blaine Fowers, Frank Richardson, and Brent Slife, "human beings do not flourish *despite* the difficulties they face. Rather, we flourish as human beings through approaching these difficulties in the best ways."[1] You could also say negative social determinants can contribute in positive ways to the quality of life when they are effectively dealt with and overcome. This is not only good news for me and my family, but for many of my former neighbors who are still languishing.

From a Christian perspective, it is also important to note that in Jesus' ministry on earth, there is a significant emphasis on physical and spiritual healing. My pastor says that it has been argued by many theologians that the whole arc of Scripture bends toward healing and restoration. As Jesus says in John 10:10, "I have come that they may have life, and have it to the full." Health equity, on this side of heaven, was and is God's desire for his family, of which we all are sons and daughters together.

Each day I receive more clarity about my calling on earth, but the crux of it is to help usher in the reality of health equity for all people, particularly those in inner-city environments. I believe this can and should be done by addressing social determinants of health and effectively casting away structures and burdens that keep people from new and unending life. Even though the path has been wild and messy, I can see God using my life for a greater purpose and that nothing has been wasted.

There is a quote from Cheryl Strayed's book *Wild: From Lost to Found on the Pacific Crest Trail* that perfectly captures how I feel about where I have come from: "I didn't feel sad or happy. I didn't feel proud or ashamed. I only felt that in spite of all the things I'd done wrong, in getting myself here, I'd done right."[2] No matter how you got here, to the final chapter of this book, I hope you will join us in this difficult work of fighting for health equity. May we band together and work thoughtfully and tirelessly to eliminate the life span and quality of life disparities in our neighborhoods because there is a better way to restore our communities.

ACKNOWLEDGMENTS

We have many to thank who have made this book possible. First, we would like to thank our colleagues, neighbors, and patients who have allowed us to be part of their lives and shaped our understanding of social determinants of health and health equity. Thank you for sharing your stories and challenging our misconceptions.

Thank you to our editor, Al Hsu, who took a chance on some first-time authors and championed this book. We would also like to thank Keri Norris, N. J. Kim, Dr. Scott Santibanez, Eric Squires, Pam Cordero, Matt Lathrop, and Cyrl Kitchens, who read each chapter and provided important feedback.

This book would not have been possible without the eager participation of community leaders and innovators whose interviews provided significant content for this book. They are:

Mariam Abdul-Aziz, CPS certification coordinator, Georgia Certified Peer Specialist Project;

Amy Becklenberg, instructor, Nell Hodgson Woodruff School of Nursing, Emory University;

Christopher Burke, director of Community and Government Relations, Georgia Institute of Technology;

Adolphus Chandler, sr. site supervisor, Georgia Works;

Michael Halicki, executive director, Park Pride;

Phillip Hunter, executive director, Georgia Works;

Odetta MacLeish-White, managing director, TransFormation Alliance;

Don Miller, counselor/therapist, Georgia Works;

Sam Rapier, CPS training coordinator, Georgia Certified Peer Specialist Project;

Treanda Smith, medical services coordinator, Good Samaritan Health Center;

Cole Thaler, Safe and Stable Homes Project director, Atlanta Volunteer Lawyers Foundation;

Sam Tsemberis, founder, Pathways Housing First;

Matias Valenzuela, director of the Office Equity and Social Justice, King County, Washington;

Dr. Bill Warren, founder and CEO, Good Samaritan Health Center;

Anthony Wilkes, community partnerships manager, Good Samaritan Health Center;

Dr. Jen, Mary Kelly, and others whose identity we have protected.

Finally, we want to thank our families. No accomplishment happens in a silo, and your love and support have been our backbone in all personal and professional endeavors.

DISCUSSION QUESTIONS

PART ONE

1. What are your thoughts on the idea that social determinants of health contribute to health status and life expectancy? Do you agree or disagree? What surprised you?
2. How have you seen social determinants impact your personal health or the health of your community?
3. If you have ever worked or volunteered in ministry, community development, or other service areas, can you relate to the feelings of failure and distress expressed in this narrative? What helped you heal? What do you think could help you heal if you are feeling that way now?
4. Was there a particular story or statistic that stood out to you? Why do you think it affected you the way it did?
5. As you think about improving your community, be it your neighborhood, local ministry, or service population, what are your major takeaways from what you have read so far? Can you identify strategies or approaches that need to change or could be improved?

PART TWO

1. The authors discuss how they confronted their personal biases and misconceptions during their work in urban Atlanta. Have you had experiences like this in your life? Did anything in this book cause you to consider biases you might hold?
2. In reading about strategies toward health equity, from Good Sam to Georgia Works to Housing First, how did you feel? Were these examples hopeful, challenging, encouraging, or overwhelming?
3. As you read through the examples of local initiatives, which were the most interesting and inspiring to you? What did you identify as the key factors contributing to their success?
4. Throughout part two and in particular in chapter eighteen, there are ideas for ways individuals can influence social determinants and contribute to health equity. Which of these ideas make sense for you? What changes, if any, are you considering as a result of what you have read?
5. Based on your experience, do you have ideas for change and improving the health of your community that are not included in this book? How could you build on these ideas to create positive change?

NOTES

1 TWO JOURNEYS TO THE INNER CITY

[1]"Mapping Life Expectancy," VCU Center on Society and Health, September 26, 2016, www.societyhealth.vcu.edu/work/the-projects/mapping-life-expectancy.html.

[2]Tracy Jan, "Redlining Was Banned 50 Years Ago. It's Still Hurting Minorities Today," *Washington Post*, March 28, 2018, www.washingtonpost.com/news/wonk/wp/2018/03/28/redlining-was-banned-50-years-ago-its-still-hurting-minorities-today/?utm_term=.6c48287fe06a.

[3]"Buckhead," Point 2 Homes, accessed April 9, 2018, www.point2homes.com/US/Neighborhood/GA/Fulton-County/Atlanta/Buckhead-Demographics.html.

[4]"Public Health Assessment and Wellness," Fulton-DeKalb Hospital Authority, accessed April 9, 2018, http://fultondekalb.ga.networkofcare.org/ph/county-indicators.aspx.

[5]Karen Pooley, "Segregation's New Geography: The Atlanta Metro Region, Race, and the Declining Prospects for Upward Mobility," *Southern Spaces*, April 15, 2015, https://southernspaces.org/2015/segregations-new-geography-atlanta-metro-region-race-and-declining-prospects-upward-mobility.

[6]VCU Center on Society and Health, "Mapping Life Expectancy."

2 WHAT IS MAKING US SICK?

[1]"Concepts of Contagion and Epidemics," in *Contagion: Historical Views of Disease and Epidemics*, Harvard University Library collection, accessed April 9, 2018, http://ocp.hul.harvard.edu/contagion/concepts.html.

[2]"Dr. Semmelweis' Biography," Semmelweis Society International, 2009, http://semmelweis.org/about/dr-semmelweis-biography.

[3]National Research Council, "A Theory of Germs," in *Science, Medicine, and Animals* (n.p.: National Academies Press, 2004), 7, www.nap.edu/read/10733/chapter/4.

[4]C. N. Trueman, "Louis Pasteur," History Learning Site, March 17, 2015, www.historylearningsite.co.uk/a-history-of-medicine/louis-pasteur.

[5]Melonie Heron, "Deaths: Leading Causes for 2014," National Vital Statistics Reports 65, no. 5 (June 2016): 1.

[6]Julia Belluz, "Sir Michael Marmot on Why All Matters Are Health Matters," *Maclean's*, August 14, 2012, www.macleans.ca/society/health/sir-michael-marmot-on-why-all-matters-are-health-matters.

[7]M. G. Marmot, Geoffrey Rose, M. Shipley, and P. J. S. Hamilton, "Employment Grade and Coronary Heart Disease in British Civil Servants," *Journal of Epidemiology and Community Health* 32, no. 4 (December 1978): 244.

[8]"About Social Determinants of Health," World Health Organization, accessed April 11, 2018, www.who.int/social_determinants/sdh_definition/en.

[9]Helen Epstein, "Ghetto Miasma; Enough to Make You Sick?," *New York Times Magazine*, October 12, 2003, www.nytimes.com/2003/10/12/magazine/ghetto-miasma-enough-to-make-you-sick.html.

[10]Bruce McEwen and Teresa Seeman, "Allostatic Load and Allostasis," in Allostatic Load Notebook, August 2009, www.macses.ucsf.edu/research/allostatic/allostatic.php.

[11]McEwen and Seeman, "Allostatic Load and Allostasis."

[12]McEwen and Seeman, "Allostatic Load and Allostasis."

[13]A. Steptoe and M. Marmot, "The Role of Psychobiological Pathways in Socio-Economic Inequalities in Cardiovascular Disease Risk," *European Heart Journal* 23, no.1 (January 2002): 15.

[14]Pathik Wadhwa, Snja Entringer, Claudia Buss, and Michael Lu, "The Contribution of Maternal Stress to Preterm Birth: Issues and Considerations," *Clinical Perinatology* 38 (September 2011); 351-84, www.ncbi.nlm.nih.gov/pmc/articles/PMC3179976/pdf/nihms318210.pdf; Steptoe and Marmot, "Role of Psychobiological Pathways in Socio-Economic Inequalities in Cardiovascular Disease Risk," 21; and Jane Ellen Clougherty, Ellen A. Eisen, Martin D. Slade, Ichiro Kawachi, and Mark R. Cullen. "Workplace Status and Risk of Hypertension Among Hourly and Salaried Aluminum Manufacturing Employees," *Social Science and Medicine* 68, no. 2 (January 2009), 304-13, www.ncbi.nlm.nih.gov/pmc/articles/PMC2659853/pdf/nihms92772.pdf.

[15]Paula Braveman and Laura Gottlieb, "The Social Determinants of Health: It's

Time to Consider the Causes of the Causes," *Public Health Reports* 129, suppl. 2 (January-February 2014): 19-31, www.ncbi.nlm.nih.gov/pmc/articles/PMC3863696.

[16]"Social Determinants of Health," HealthyPeople.gov, last modified April 13, 2018, www.healthypeople.gov/2020/topics-objectives/topic/social-determinants-of-health.

[17]Breanna and Veronica, while employees at the Good Samaritan Health Center, contributed to this book on their own time and in their personal capacity. The views expressed are their own and do not necessarily represent the views of the Good Samaritan Health Center. Any religious views shared in this book are the authors' and do not reflect the beliefs and opinions of individuals who were interviewed or otherwise contributed to this book.

[18]Paula Braveman, Elaine Arkin, Tracy Orleans, Dwayne Proctor, and Alonzo Plough, "What Is Health Equity? And What Difference Does a Definition Make?" Robert Wood Johnson Foundation, May 1, 2017, www.rwjf.org/en/library/research/2017/05/what-is-health-equity-.html.

[19]"A Christian Critique of Equity," Christian Community Health Fellowship, January 1, 2005, www.cchf.org/resources/h-and-da-christian-critique-of-equity.

3 THE TRAUMA OF POVERTY

[1]Eric Jenson, "The Effects of Poverty on the Brain," Science Network, accessed April 23, 2018, http://thesciencenetwork.org/docs/BrainsRUs/Effetcs%20of%20Poverty_Jensen.pdf.

[2]"Why Place Matters: Building the Movement for Healthy Communities," PolicyLink, 2007, www.policylink.org/find-resources/library/why-place-matters-building-the-movement-for-healthy-communities.

[3]Richard Bentall, "Mental Illness Is a Result of Misery, yet We Still Stigmatize It," Foundation for Excellence in Mental Health, February 26, 2016, www.mentalhealthexcellence.org/mental-illness-is-a-result-of-misery-yet-still-we-stigmatise-it.

[4]"Poverty Rate by Race/Ethnicity," Kaiser Family Foundation, accessed April 23, 2018, www.kff.org/other/state-indicator/poverty-rate-by-raceethnicity.

[5]PolicyLink, "Why Place Matters."

[6]"Toxic Stress," Harvard University Center on the Developing Child, accessed April 23, 2018, http://developingchild.harvard.edu/science/key-concepts/toxic-stress.

[7]Abraham Maslow, quoted in Saul McLeod, "Maslow's Hierarchy of Needs," Simply Psychology, updated May 2018, www.simplypsychology.org/maslow.html.

[8]Wess Stafford, *Too Small to Ignore* (Colorado Springs, CO: WaterBrook Press, 2005), 184.

[9]Lisa Signorello, Sarah Cohen, David Williams, and Heather Munro, Margaret Hargreaves, and William Blot, "Socioeconomic Status, Race, and Mortality: A Prospective Cohort Study," *American Journal of Public Health* 104, no. 12 (October 2014): e102.

[10]Yang Jiang, Maribel R. Granja, and Heather Koball, "Basic Facts About Low-Income Children, Children under 18 Years, 2015," National Center for Children in Poverty, January 2017, www.nccp.org/publications/pub_1170.html.

[11]Michael Marmot and Richard G. Wilkinson, *Social Determinants of Health,* 2nd ed. (New York: Oxford University Press, 2006), chap 3.

[12]"Overcoming Obstacles to Health: Report from the Robert Wood Johnson Foundation to the Commission to Build a Healthier America," Robert Wood Johnson Foundation, February 2008, www.commissiononhealth.org/PDF/ObstaclesToHealth-Report.pdf.

[13]Sheldon Cohen, Denise Janicki-Deverts, Edith Chen, and Karen Matthews, "Childhood Socioeconomic Status and Adult Health," *Annals of the New York Academy of Sciences* 1186 (February 16, 2010), 37-55, https://nyaspubs.onlinelibrary.wiley.com/doi/full/10.1111/j.1749-6632.2009.05334.x.

[14]Dr. Jen, personal interview, August 23, 2017. Additional references to Dr. Jen are taken from this interview.

[15]"Poverty Rate by Race/Ethnicity," Kaiser Family Foundation.

[16]David Williams and Chiquita Collins, "Racial Residential Segregation: A Fundamental Cause of Racial Disparities in Health," *Public Health Reports* 116 (September-October 2001), 405.

[17]Williams and Collins, "Racial Residential Segregation," 405.

[18]Yin Paradies, "A Systematic Review of Empirical Research on Self-Reported Racism and Health," *International Journal of Epidemiology* 35, no. 4: 895.

[19]Bryan Stevenson, introduction to *Just Mercy: A Story of Justice and Redemption* (New York: Random House, 2004).

4 WORKING TO DEATH

[1]Lecrae, "Just Like You," *Rehab*, Reach Records, September 28, 2010.

[2]Phillip Hunter, personal interview, August 1, 2018. Additional references to Phillip Hunter are taken from this interview.

[3]S. Jay Olshansky et al., "Differences in Life Expectancy Due to Race and Educational Differences Are Widening, and May Not Catch Up," *Health Affairs* 31, no. 8 (August 2012): 1805.

[4]Anne Driscoll and Amy Bernstein, "Health and Access to Care Among Employed and Unemployed Adults: United States, 2009-2010," *NCHS Data Brief* 83 (January 2012), www.cdc.gov/nchs/data/databriefs/db83.pdf; Jennifer Pharr, Sheniz Moonie, and Timothy Bungum, "The Impact of Unemployment on Mental and Physical Health, Access to Health Care and Health Risk Behaviors," *ISRN Public Health*, October 2011, www.hindawi .com/journals/isrn/2012/483432; and Frances McKee-Ryan, Zhaoli Song, Connie Wanberg, and Angelo Kinicki, "Psychological and Physical Well-Being During Unemployment: A Meta-Analytic Study," *Journal of Applied Psychology* 90, no. 1 (January 2005): 67.

[5]Driscoll and Bernstein, "Health and Access to Care"; and Pharr, Moonie, and Bungum, "Impact of Unemployment."

[6]"Out Of Reach 2018," National Low Income Housing Coalition, accessed April 23, 2018, http://nlihc.org/oor.

[7]"Key Facts About the Uninsured Population," Kaiser Family Foundation, updated November 29, 2017, www.kff.org/uninsured/fact-sheet/key-facts -about-the-uninsured-population.

[8]M. G. Marmot, Geoffrey Rose, M. Shipley, and P. J. S. Hamilton, "Employment Grade and Coronary Heart Disease in British Civil Servants," *Journal of Epidemiology and Community Health* 32, no. 4 (December 1978): 244.

[9]Jane Clougherty, Kerry Souza, and Mark Cullen, "Work and Its Role in Shaping the Social Gradient in Health," *Annals of the New York Academy of Sciences* 1186 (February 2010): 106.

[10]Oliver Hammig and Georg Bauer, "The Social Gradient in Work and Health: A Cross-Sectional Study Exploring the Relationship Between Working Conditions and Health Inequalities," *BMC Public Health* 13, no. 1170 (December 2013), https://bmcpublichealth.biomedcentral.com/articles/10.1186 /1471 -2458-13-1170.

[11]Michael Marmot and Richard G. Wilkinson, *Social Determinants of Health*, 2nd ed. (New York: Oxford University Press, 2006), chap 6.

[12]Marmot and Wilkinson, *Social Determinants of Health*, chap 6. Robert Wood Johnson Foundation, "Issue Brief 4: Work and Health," December 2008, www.commissiononhealth.org/PDF/0e8ca13d-6fb8-451d-bac8-7d15 343aacff/Issue%20Brief%204%20Dec%2008%20-%20Work%20and%20

Health.pdf; and Mika Kivimaki et al., "Job Strain as a Risk Factor for Coronary Heart Disease: A Collaborative Meta-Analysis of Individual Participant Data," *Lancet* 380 (October 2012), 1491-97, www.ncbi.nlm.nih.gov/pmc/articles /PMC3486012.

[13]"Why Place Matters: Building the Movement for Healthy Communities," PolicyLink, 2007, www.policylink.org/find-resources/library/why-place -matters-building-the-movement-for-healthy-communities.

[14]Jane Clougherty, Kerry Souza, and Mark Cullen, "Work and Its Role in Shaping the Social Gradient in Health," *Annals of the New York Academy of Sciences*, 1186 (February 2010): 106.

[15]Annika Rosengren et al., "Association of Psychosocial Risk Factors with Risk of Acute Myocardial Infarction in 11119 Cases and 13648 Controls from 52 Countries (the INTERHEART study): Case-Control Study," *Lancet,* 364, no. 9438 (September 2004): 953-62, www.thelancet.com/journals/lancet/article /PIIS0140-6736(04)17019-0/fulltext.

5 KOOL-AID IN A BABY BOTTLE

[1]David Williams and Chiquita Collins, "Racial Residential Segregation: A Fundamental Cause of Racial Disparities in Health," *Public Health Reports* 116 (September-October 2001): 410.

[2]"Hunger and Poverty Facts," Feeding America, accessed April 23, 2018, www .feedingamerica.org/hunger-in-america/hunger-and-poverty-facts.html.

[3]Daniel Jones, "Unhealthy Food Advertising Targets Black and Hispanic Youth," *UCONN Today*, August 11, 2015, https://today.uconn.edu/2015/08 /unhealthy-food-advertising-targets-black-and-hispanic-youth.

[4]Christopher Burke, personal interview, August 26, 2017. All additional references to Christopher Burke were taken from this interview.

[5]"Overcoming Obstacles to Health," Robert Wood Johnson Foundation, February 2008, www.commissiononhealth.org/PDF/ObstaclesToHealth -Report.pdf.

[6]"Why Education Matters to Health: Exploring the Causes," VCU Center on Society and Health, February 13, 2015, http://societyhealth.vcu.edu /work/the-projects/why-education-matters-to-health-exploring-the -causes.html.

[7]Helen Epstein, "Ghetto Miasma: Enough to Make You Sick?" *New York Times Magazine,* October 12, 2003.

6 LONGEVITY AND LEARNING

[1]Amy Schulz et al., "Unfair Treatment, Neighborhood Effects, and Mental Health in the Detroit Metropolitan Area," *Journal of Health and Social Behavior* 41, no. 3 (September 2000): 320.

[2]David Williams and Chiquita Collins, "Racial Residential Segregation: A Fundamental Cause of Racial Disparities in Health," *Public Health Reports* 116 (September-October 2001): 406.

[3]Williams and Collins, "Racial Residential Segregation," 406.

[4]"Why Does Education Matter So Much to Health?," VCU Center on Society and Health, February 13, 2015, www.rwjf.org/content/dam/farm/reports/issue_briefs/2012/rwjf403347.

[5]"How Do Neighborhoods Affect Economic Opportunity," Equity Opportunity Project, accessed April 23, 2018, www.equality-of-opportunity.org/neighborhoods.

[6]"Health, United States, 2011: With Special Feature on Socioeconomic Status and Health," National Center for Health Statistics, May 2012, www.cdc.gov/nchs/data/hus/hus11.pdf.

[7]S. Jay Olshansky et al., "Differences in Life Expectancy Due to Race and Educational Differences Are Widening, and May Not Catch Up," *Health Affairs* 31, no. 8 (August 2012): 1805.

[8]"Why Education Matters to Health: Exploring the Causes," VCU Center on Society and Health, February 13, 2015, http://societyhealth.vcu.edu/work/the-projects/why-education-matters-to-health-exploring-the-causes.html.

[9]Ahmed Jamal et al., "Current Cigarette Smoking Among Adults—United States, 2005–2015," *MMWR*, 65, no. 44 (November 11, 2016), www.cdc.gov/mmwr/volumes/65/wr/mm6544a2.htm?s_cid=mm6544a2_w.

[10]VCU Center on Society and Health, "Why Does Education Matter to Health?"

[11]J. S. Schiller, J. W. Lucas, and J. A. Peregoy, "Summary Health Statistics for U.S. Adults: National Health Interview Survey, 2011," *Vital Health Stat* 10, no. 256 (December 2012), www.cdc.gov/nchs/data/series/sr_10/sr10_256.pdf.

[12]Helen Epstein, "Ghetto Miasma; Enough to Make You Sick?," *New York Times Magazine,* October 12, 2003.

[13]Sally Banks Zakariva, "Learning to Read, Reading to Learn: Why Third Grade Is a Pivotal Year for Mastering Literacy," Center for Public Education. March 2015, www.nsba.org/sites/default/files/reports/NSBA_CPE_Early_Literacy_Layout_2015.pdf.

[14]Joy Lesnick, Robert George, Cheryl Smithgall, and Julia Gwynne, "Reading on Grade Level in Third Grade: How Is It Related to High School Performance and College Enrollment?," Chapin Hall at the University of Chicago, 2010, www.aecf.org/m/resourcedoc/aecf-ReadingonGradeLevelLongAnal-2010.PDF.

[15]Pascael Beaudette, Kanti Chalasani, and Sam Rauschenberg, "How Do Students' 3rd Grade Reading Levels Relate to Their ACT/SAT Performance and Chance of Graduating from High School?" March 22, 2017, https://gosa.georgia.gov/sites/gosa.georgia.gov/files/related_files/press_release/3rd%20Grade%20Reading-Graduation-ACT-SAT%20Analysis%20Final%2003222017.pdf.

[16]Anne Fernald, Virginia Marchman, and Adriana Weisleder, "SES Differences in Language Processing Skill and Vocabulary Are Evident at 18 Months," *Developmental Science* 16, no. 2 (March 2013): 234-48, www.ncbi.nlm.nih.gov/pmc/articles/PMC3582035.

[17]Betty Hart and Todd R. Risley, "The Early Catastrophe: The 30 Million Word Gap by Age 3," *American Educator* (Spring 2003), www.aft.org/sites/default/files/periodicals/TheEarlyCatastrophe.pdf.

[18]Christine Duchouquette, Kristen Loschert, and Patte Barth, "Beyond Fiction: The Importance of Reading for Information," NSBA Center for Public Education, October 2014, https://cdn-files.nsba.org/s3fs-public/Beyond-Fiction-Full-Report-PDF.pdf.

[19]Duchouquette, Loschert, and Barth, "Beyond Fiction."

7 WHEN HOUSING HURTS

[1]David Williams and Chiquita Collins, "Racial Residential Segregation: A Fundamental Cause of Racial Disparities in Health," *Public Health Reports* 116 (September-October 2001): 409

[2]"Why Place Matters: Building the Movement for Healthy Communities," PolicyLink, 2007, www.policylink.org/find-resources/library/why-place-matters-building-the-movement-for-healthy-communities.

[3]PolicyLink, "Why Place Matters."

[4]PolicyLink, "Why Place Matters."

[5]Cole Thaler, personal interview, July 18, 2017. Additional references to Cole Thaler are taken from this interview.

[6]"Our Mission," Atlanta Volunteer Lawyers Foundation, accessed April 23, 2018, https://avlf.org.

[7]"Affordable Housing," U.S. Department of Housing and Urban Development, accessed April 23, 2018 https://portal.hud.gov/hudportal/HUD?src=/program_offices/comm_planning/affordablehousing.

[8]"HUD Modifies Extremely-Low Income Definition," National Housing and Rehabilitation Association, July 9, 2014, www.housingonline.com/2014/07/09/hud-modifies-extremely-low-income-definition.

[9]Josh Leopold, Liza Getsinger, Pamela Blumenthal, Katya Abazajian, and Reed Jordan, "The Housing Affordability Gap for Extremely Low Income Renters in 2013," June 2015, www.urban.org/sites/default/files/publication/54106/2000260-The-Housing-Affordability-Gap-for-Extremely-Low-Income-Renters-2013.pdf.

[10]National Law Center on Poverty and Homelessness. "Homelessness in America: Overview of Data and Causes," updated January 2015, www.nlchp.org/documents/Homeless_Stats_Fact_Sheet.

[11]"Mapping America's Rental Housing Crisis," Urban Institute, updated April 17, 2017, http://apps.urban.org/features/rental-housing-crisis-map.

[12]Deirdre Oakley, Erin Ruel, and Lesley Reid, "'It Was Really Hard . . . It Was Alright . . . It Was Easy,' Public Housing Relocation Experiences and Destination Satisfaction in Atlanta," *Cityscape* 15, no. 2 (2013): 181.

[13]Treanda Smith, personal interview, July 12, 2017. Additional references to Treanda Smith are taken from this interview.

[14]Griff Tester, Erin Ruel, Angela Anderson, Donald Reitzes, and Deirdre Oakley, "Sense of Place Among Atlanta Public Housing Residents," *Journal of Urban Health* 88, no. 3 (June 2001): 436-53, www.ncbi.nlm.nih.gov/pmc/articles/PMC3126922.

[15]Oakley, Ruel, and Reid, "It Was Really Hard," 181.

[16]Otis Rolley, "TEDxMidAtlantic 2010—Otis Rolley," November 5, 2010, www.youtube.com/watch?v=rfka3clhZLUandt=10s.

[17]Robert Bullard, "Urban Infrastructure: Social Environmental, and Health Risks to African Americans," in *Handbook of Black American Health: The Mosaic of Conditions, Issues, Policies and Prospects*, ed. Ivor Lensworth Livingston (Westport, CT: Greenwood, 1994), 315-30.

8 THE CHALLENGE OF GETTING WELL

[1]"The Uninsured a Primer 2013-4: How Does Lack of Insurance Affect Access to Health Care?," Kaiser Family Foundation, November 14, 2003, www.kff.org/report-section/the-uninsured-a-primer-2013-4-how-does-lack-of-insurance-affect-access-to-health-care.

[2]Kaiser Family Foundation, "Uninsured a Primer 2013-4."

[3]Sara R. Collins, Michelle M. Doty, Ruth Robertson, and Tracy Garber. "Help on the Horizon: How the Recession Has Left Millions of Workers Without Health Insurance, and How Health Reform Will Bring Relief," Commonwealth Fund, March 16, 2011, www.commonwealthfund.org /publications/fund-reports/2011/mar/help-on-the-horizon.

[4]Andrew Wilper, Steffie Woolhandler, Karen Lasser, Danny McCormick, Daivd Bor, and David Himmelstein, "Health Insurance and Mortality in US Adults," *American Journal of Public Health* 99, no. 12 (December, 2009): 2292.

[5]Fizan Abdullah et al., "Analysis of 23 Million US Hospitalizations: Uninsured Children Have Higher All-Cause In-Hospital Mortality," *Journal of Public Health* 32, no. 2 (June 2010): 236-44, https://academic.oup.com /jpubhealth/article/32/2/236/1607607.

[6]Collins, Doty, Robertson, and Garber, "Help on the Horizon."

[7]Silvia Tejada, Julie Darnell, Young Cho, Melinda Stolley, Talar Markossian, and Elizabeth Calhoun, "Patient Barriers to Follow-Up Care for Breast and Cervical Cancer Abnormalities," *Journal of Women's Health* 22, no. 6 (June 2013): 510.

[8]Arica White et al., "Cancer Screening Test Use—United States, 2015," *MMWR Morbidity and Mortality Weekly Report* 66, no. 8 (March 2017): 201-6, http://dx.doi.org/10.15585/mmwr.mm6608a1.

[9]Stacey McMorrow, Genevieve Kenney, and Dana Goin, "Determinants of Receipt of Recommended Preventive Services: Implications for the Affordable Care Act," *American Journal of Public Health* 104, no. 12 (December 2014): 2394-96.

[10]"Projecting the Supply and Demand for Primary Care Practitioners Through 2020," Health Resources and Services Administration, November 2013, https://bhw.hrsa.gov/sites/default/files/bhw/nchwa/projectingprimarycare.pdf.

[11]Health Resources and Services Administration, "Projecting the Supply and Demand."

[12]"Designated Health Professional Shortage Areas Statistics," HRSA Data Warehouse, accessed August 26, 2017, https://datawarehouse.hrsa.gov /tools/analyzers/hpsafind.aspx.

[13]"Health Disparities in Rural Women," American College of Obstetrics and Gynecologists, February 2014, www.acog.org/-/media/Committee-Opinions/Committee-on-Health-Care-for-Underserved-Women/co586.pdf?dmc=1ands=20170827T0011234491.

[14]"Health Expenditure," Organization for Economic Cooperation and Development, accessed April 27, 2018, www.oecd.org/els/health-systems/health-expenditure.htm.

[15]James Macinko, Barbara Starfield, and Leiyu Shi, "The Contribution of Primary Care Systems to Health Outcomes Within Organization for Economic Cooperation and Development (OECD) Countries, 1970–1998," *Health Services Research* 38, no. 3 (May 2003): 858.

[16]"Mental Illness," National Institute of Mental Health, updated November, 2017, www.nimh.nih.gov/health/statistics/prevalence/any-mental-illness-ami-among-us-adults.shtml.

[17]"The Psychiatric Shortage: Causes and Solutions," National Council for Behavioral Health, March 28, 2017, www.thenationalcouncil.org/wp-content/uploads/2017/03/Psychiatric-Shortage_National-Council-.pdf.

[18]Sam Rapier and Mariam Abdul-Aziz, personal interview, July 27, 2017.

[19]"10 Leading Causes of Death by Age Group United States-2014," Center for Disease Control and Prevention, accessed April 27, 2018, www.cdc.gov/injury/images/lc-charts/leading_causes_of_death_age_group_2014_1050w760h.gif.

10 A NEW BEGINNING

[1]"Why Place Matters: Building the Movement for Healthy Communities," PolicyLink, 2007, www.policylink.org/find-resources/library/why-place-matters-building-the-movement-for-healthy-communities.

[2]"Welcome to the Avielle Foundation," Avielle Foundation, accessed April 27, 2018, https://aviellefoundation.org.

[3]"Our History and Founders," Enterprise Community Partners," accessed April 27, 2018, www.enterprisecommunity.org/about/what-we-do/our-history.

[4]Kristian Foden-Vencil, "How Doctors Find Value in Knowing Your Socioeconomic Data," *OPB*, July 12, 2017, www.opb.org/news/article/health-care-social-data-information-economics-kaiser-permanente.

[5]"The Collective Impact Framework," Collaboration for Impact, accessed April 27, 2018, www.collaborationforimpact.com/collective-impact.

11 THE GOOD SAM STORY

[1]"Addressing the Life Expectancy Gap: The Good Samaritan Full Circle of Health," Good Samaritan Health Center, September 2016, http://goodsamatlanta.org/wp-content/uploads/2017/02/Full-Circle-of-Health-White-Paper.pdf.

[2]Dr. Bill Warren, personal interview, August 10, 2017. Additional references to Bill Warren are taken from this interview.

[3]Phaedra Corso, Rebecca Walcott, and Justin Ingels, "Return on Investment in Georgia Charitable Care Network (GCCN) Clinics," Economic Evaluation Research Group, October 12, 2015, http://charitablecarenetwork.com/wp-content/uploads/2012/02/EERG-report_10-12-2015.pdf.

[4]"America's Free and Charitable Clinics: Vital Support for 30 Million Uninsured Americans," National Association of Free and Charitable Clinics, accessed April 27, 2018, www.communityhealth.org/wp-content/uploads/2012/03/Americas-Free-and-Charitable-Clinics-NAFCC_Report_081414_-2.pdf.

[5]Dan Pallotta, "'Efficiency' Measures Miss the Point,'" *Harvard Business Review*, June 22, 2009, https://hbr.org/2009/06/efficiency-measures-miss-the-p.

[6]Scott Santibanez and Veronica Squires, "Balancing Hope and Critical Analysis in Community Service," Emerging Scholars Blog, accessed April 22, 2018, http://blog.emergingscholars.org.

12 POVERTY

[1]Mary Kelly, personal interview, November 7, 2017. Additional references to Mary Kelly are taken from this interview.

[2]Michael Meit, Megan Heffernan, Erin Tanenbaum, and Topher Hoffmann, "Appalachian Diseases of Despair," Appalachian Regional Commission, August 2017, www.arc.gov/assets/research_reports/AppalachianDiseasesofDespairAugust2017.pdf.

[3]Elizabeth Kneebone, "The Growth and Spread of Concentrated Poverty, 2000 to 2008–2012," Brookings, July 31, 2014, www.brookings.edu/interactives/the-growth-and-spread-of-concentrated-poverty-2000-to-2008-2012.

[4]Wilmot Allen, "Urban Poverty in America: The Truly Disadvantaged Revisited," *HuffPost*, July 25, 2014, www.huffingtonpost.com/wilmot-allen/urban-poverty-in-america_b_2516832.html.

[5]Patrick Sharkey, "Ending Urban Poverty: The Inherited Ghetto," *Boston Review*, January 1, 2008, http://bostonreview.net/patrick-sharkey-inherited-ghetto-racial-inequality.

[6]Sharkey, "Ending Urban Poverty."

[7]Catherine Cozzarelli, Anna Wilkinson, and Michael Tagler, "Attitudes Toward the Poor and Attributions for Poverty," *Journal of Social Issues* 57, no. 2 (December 2002): 215.

[8]Odetta MacLeish-White, personal interview, August 10, 2017. All additional references to Odetta MacLeish-White refer to this interview.

[9]Sam Tsemberis, personal interview, November 7, 2017. Additional references to Sam Tsemberis are taken from this interview.

[10]"King County Equity and Social Justice Annual Report," Office of Equity and Social Justice, December, 2015, www.kingcounty.gov/~/media/elected/executive/equity-social-justice/2015/2015_ESJ_Report.ashx?la=en.

[11]Matias Valenzuela, personal interview, October 20, 2017. Additional references to Matias Valenzuela are taken from this interview.

[12]Valenzuela, personal interview.

[13]Abigail Beatty and Dionne Foster, "The Determinants of Equity," King County Office of Performance, Strategy and Budget, January 2015, www.kingcounty.gov/~/media/elected/executive/equity-social-justice/2015/The_Determinants_of_Equity_Report.ashx?la=en.

[14]"2015 Equity Impact Review Process Overview," King County, updated March 2016, www.kingcounty.gov/~/media/elected/executive/equity-social-justice/2016/The_Equity_Impact_Review_checklist_Mar2016.ashx?la=en.

13 HIRED AND HEALTHY

[1]Jane Clougherty, Kerry Souza, and Mark Cullen, "Work and Its Role in Shaping the Social Gradient in Health," *Annals of the New York Academy of Sciences* 1186 (February 2010): 102.

[2]Frances McKee-Ryan, Zhaoli Song, Connie Wanberg, and Angelo Kinicki, "Psychological and Physical Well-Being During Unemployment: A Meta-Analytic Study," *Journal of Applied Psychology* 90, no. 1 (January 2005): 54.

[3]Susan Egerter, Jane An, Rebecca Grossman-Kahn, and Paula Braveman, "Issue Brief 4: Work and Health," Robert Wood Johnson Foundation, December 2008, http://www.commissiononhealth.org/PDF/0e8ca13d-6fb8-451d-bac8-7d15343aacff/Issue%20Brief%204%20Dec%2008%20-%20Work%20and%20Health.pdf.

[4]"The Georgia Works! Program," Georgia Works, accessed April 29, 2018, www.georgiaworks.net/about-1.

[5]Phillip Hunter, Don Miller, and Adolphus Chandler, personal interview, August 1, 2018. Additional references to Georgia Works are taken from this interview.

[6]Kyle Wingfield, "Program for Atlanta's Homeless Men Shows Work Really Works," *Atlanta Journal-Constitution,* October 12, 2016, www.myajc.com/news/opinion/program-for-atlanta-homeless-men-shows-work-really-works/ART58yrfpY1Cg4073Ql1OL.

[7]"2016 State Statistics Information," National Institute of Corrections, accessed April 29, 2018, https://nicic.gov/state-statistics-information.

[8]Tamara Easterlin, Farnaz Jafari, Sumaiya Khan, and Steven Parente, "Cost-Benefit Analysis of the Georgia Works! Program to End Chronic Homelessness in Atlanta," Georgia Institute of Technology, July 27, 2016.

[9]Omid Elyaskordi, Melanie McNeely, Wesley Sanders, and Stephen Steinmann, "Georgia Works! Program Legal and Healthcare Cost Savings for Ending Chronic Homelessness in Atlanta," Georgia Institute of Technology, July 27, 2016.

14 GROWING HEALTH FROM THE GROUND UP

[1]"Nutrition, Physical Activity, and Obesity," HealthyPeople.gov, accessed April 29, 2018, www.healthypeople.gov/2020/leading-health-indicators/2020-lhi-topics/Nutrition-Physical-Activity-and-Obesity.

[2]James Levine, "Poverty and Obesity in the U.S.," *Diabetes* 60 no. 11 (November 2011): 2667.

[3]David Williams and Chiquita Collins, "Racial Residential Segregation: A Fundamental Cause of Racial Disparities in Health," *Public Health Reports* 116 (September-October 2001): 411.

[4]Breanna Lathrop and Ursula Pritham, "A Pilot Study of Prenatal Care Visits Blended Group and Individual for Low Income Women," *Nursing for Women's Health* 18, no. 6 (December 2014): 466.

[5]"Atlanta's Local Food Baseline Report," Food Well Alliance, fall 2017, https://static1.squarespace.com/static/543c2e74e4b0a10347055c4d/t/59d6abd58a02c7c9b996ac07/1507240935571/Food+Well+Alliance_Local+Food+Baseline+Report+2017_Final.pdf.

[6]Food Well Alliance, "Atlanta's Local Food Baseline Report."

[7]Anthony Wilkes, personal interview, August 9, 2017.

15 A HEALTHY START

[1]Dominque Jordan Turner, "Poverty Is a SuperPower," TEDx, June 10, 2017, www.youtube.com/watch?v=LyVFL3iGRKM.

[2]"Child Poverty," National Center for Children in Poverty, accessed May 4, 2018, www.nccp.org/topics/childpoverty.html.

[3]"Word Gap," Talk With Me Baby, accessed May 4, 2018, www.talkwith mebaby.org.

[4]Betty Hart and Todd Risley, "The Early Catastrophe: The 30 Million Word Gap by Age 3," *American Educator* (spring 2003), www.aft.org/sites/default /files/periodicals/TheEarlyCatastrophe.pdf.

[5]Amy Becklenberg, personal interview, July 20, 2017. All additional references to Amy Becklenberg are take from this interview.

[6]Amy Rothschild, "Beyond the Word Gap" *Atlantic*, April 22, 2016, www .theatlantic.com/education/archive/2016/04/beyond-the-word-gap /479448.

[7]Anne Fernald, Virginia Marchman, and Adriana Weisleder, "SES Differences in Language Processing Skill and Vocabulary Are Evident at 18 Months," *Developmental Science,* 16, no. 2 (March 2013): 234-48, www .ncbi.nlm.nih.gov/pmc/articles/PMC3582035.

[8]Netta Avineri and Eric Johnson, introduction to "Invited Forum: Bridging the 'Language Gap,'" *Journal of Linguistic Anthropology* 25, no. 1 (2015): 67-68.

[9]Shelia Brice Health, "The Simple and Direct? Almost Never the Solution," in "Invited Forum: Bridging the 'Language Gap,'" 68.

[10]Coleen Boyle et al., "Trends in the Prevalence of Developmental Disabilities in US Children, 1997–2008," *Pediatrics* 127, no. 6 (June 2011): 1037.

16 A PLACE TO CALL HOME

[1]Pathways Housing First, accessed May 4, 2018, www.pathwayshousing first.org.

[2]Deborah Padgett, Benjamin Henwood, and Sam Tsemberis, *Housing First: Ending Homelessness, Transforming Systems, and Changing Lives* (New York, NY: Oxford University Press, 2016), chap. 1.

[3]Leyla Gulcur, Ana Stefancic, Marybeth Shinn, Sam Tsemberis, and Sean Fischer, "Housing, Hospitalizations, and Cost Outcomes for Homeless Individuals with Psychiatric Disabilities Participating in Continuum of Care and Housing First Programmes," *Journal of Community and Applied Social Psychology* 13 (2003): 177, 179.

[4]Sam Tsemberis, Levla Gulcur, and Maria Nakae, "Housing First, Consumer Choice, and Harm Reduction for Homeless Individuals with a Dual Diagnosis," *American Journal of Public Health* 94, no. 4 (April 2004): 654.

[5]Padgett, Henwood, and Tsemberis, *Housing First*, chap. 4.

[6]Tama Leventhal and Jeanne Brooks-Gunn, "Moving to Opportunity: An Experimental Study of Neighborhood Effects on Mental Health," *American Journal of Public Health* 93, no. 9 (September 2003): 1580.

[7]Kriston Capps, "New York City Guarantees a Lawyer to Every Resident Facing Eviction.," *CityLab*, August 14, 2017, www.citylab.com/equity/2017/08/nyc-ensures-eviction-lawyer-for-every-tenant/536508.

[8]"Housing Right to Counsel Project," DC Bar, accessed May 4, 2018, https://www.dcbar.org/pro-bono/about-the-center/right-to-counsel-project.cfm.

[9]Martin Austermuhle, "Need a Lawyer to Fight an Eviction? A New D.C. Program Provides One For Free," *WAMU American University Radio*, May 18, 2017, https://wamu.org/story/17/05/18/need-lawyer-fight-eviction-new-d-c-program-provide-one-free.

[10]Michael Halicki, personal interview, August 30, 2018. All additional references to Michael Halicki are taken from this interview.

[11]D. Merriam, A. Bality, J. Stein, and T. Boehmer, "Improving Public Health Through Public Parks and Trails: Eight Common Measures," Centers for Disease Control and Prevention, and National Park Service, 2017, http://go.nps.gov/improving_public_health.

[12]Merriam, Bality, Stein, and Boehmer, "Improving Public Health Through Public Parks and Trails."

[13]Jim Erickson, "Targeting Minority, Low-Income Neighborhoods for Hazardous Waste Sites," *Michigan News*, January 19, 2016, http://ns.umich.edu/new/releases/23414-targeting-minority-low-income-neighborhoods-for-hazardous-waste-sites.

[14]Paula Braveman, Elaine Arkin, Tracy Orleans, Dwayne Proctor, and Alonzo Plough, "What Is Health Equity? And What Difference Does a Definition Make?" May, 2017, www.rwjf.org/content/dam/farm/reports/issue_briefs/2017/rwjf437393.

[15]Mandy Thompson Fullilove, *Urban Alchemy: Restoring Joy in American's Sorted-Out Cities* (New York: New Village Press, 2013), 1-3.

[16]"New Urban Leadership Council Aims to Make Economic Case for Better Cities," New Climate Economy, November 2017, http://newclimateeconomy.net/content/release-new-urban-leadership-council-aims-make-economic-case-better-cities.

17 REENVISIONING HEALTH CARE

[1]"Understanding Health Concerns and Barriers to Accessing Care Among Underserved Populations," Health Outreach Partners, November 1, 2013, https://outreach-partners.org/2013/11/01/understanding-health-concerns-and-barriers-to-accessing-care-among-underserved-populations.

[2]Health Outreach Partners, "Understanding Health Concerns and Barriers."

[3]"Notice PIH 2013-15," Department of Housing and Urban Development, June 10, 2013, www.hud.gov/sites/documents/PIH2013-15.PDF.

[4]"The 2016 Annual Homeless Assessment Report to Congress," HUD, November 2016, www.hudexchange.info/resources/documents/2016-AHAR-Part-1.pdf.

[5]"Homeless Point-In-Time Count," Partners for Home, 2016, www.atlantaga.gov/home/showdocument?id=24072.

[6]"The U.S. Conference of Mayors Report on Hunger and Homelessness: A Status Report on Homelessness and Hunger in America's Cities," United Conference of Mayors, December 2016, https://endhomelessness.atavist.com/mayorsreport2016.

[7]"Homelessness and Health: What's the Connection?" National Healthcare for the Homeless Council, June 2011, www.nhchc.org/wp-content/uploads/2011/09/Hln_health_factsheet_Jan10.pdf; and "Why Are People Homeless?" National Alliance to End Homelessness, July 2009, www.nationalhomeless.org/factsheets/why.html.

[8]Sekene Badiaga, Didier Raoult, and Philippe Brouqui. "Preventing and Controlling Emerging and Reemerging Transmissible Diseases in the Homeless," *Emerging Infectious Diseases* 14, no. 9 (September 2008): 1354.

[9]National Alliance to End Homelessness, "Why Are People Homeless?"

[10]"Georgia Certified Peer Specialist Project," Georgia CPS Project, accessed May 5, 2018, www.gacps.org/Home.html.

[11]Sam Rapier, personal interview, July 27, 2017. All additional references to Sam Rapier are taken from this interview.

[12]Mariam Abdul-Aziz, personal interview, July 27, 2017. All additional references to Mariam Abdul-Aziz are take from this interview.

[13]"Key Facts About the Uninsured Population," Kaiser Family Foundation, updated November 9, 2017, www.kff.org/uninsured/fact-sheet/key-facts-about-the-uninsured-population.

[14]"America's Free and Charitable Clinics: Vital Support for 30 Million Uninsured Americans," National Association of Free and Charitable Clinics, August 2014, www.nafcclinics.org/sites/default/files/NAFC%20Info graphic%20%26%20Report%20Feb%202%202015%20small.pdf.

[15]"Federally Qualified Health Centers," Health Resources and Service Administration, updated March 2018, www.hrsa.gov/opa/eligibility-and -registration/health-centers/fqhc/index.html.

[16]"America's Health Centers," National Association of Community Health Centers, October 2017, www.nachc.org/wp-content/uploads/2017/11 /Americas_Health_Centers_Nov_2017.pdf.

[17]"Health at a Glance 2015: How does the United States Compare?" Organisation for Economic Co-operation and Development, accessed May 5, 2018, www.oecd.org/unitedstates/Health-at-a-Glance-2015-Key-Findings -UNITED-STATES.pdf.

18 RX FOR CHANGE

[1]Amy Schulz and Mary Northridge, "Social Determinants of Health: Implications for Environmental Health Promotion," *Health Education and Behavior* 31 no. 4 (August 2004): 456-57.

[2]"A Window of Opportunity II: An Analysis of Public Opinion on Policy," Opportunity Agenda, 2016, https://opportunityagenda.org/sites/default /files/2017-04/FINAL%20REPPORT_Updated.pdf.

[3]The Opportunity Agenda, "A Window of Opportunity II."

[4]David Williams and Chiquita Collins, "Racial Residential Segregation: A Fundamental Cause of Racial Disparities in Health," *Public Health Reports* 116 (September-October 2001): 409.

[5]Schulz and Northridge, "Social Determinants of Health," 459.

[6]Scott Santibanez, "Social Media Rants Won't Change the World but Here's How You Can," *Relevant*, October 25, 2016, https://relevantmagazine .com/reject-apathy/social-media-rants-wont-change-world.

EPILOGUE

[1]Blaine Fowers, Frank Richardson, and Brent Slife, *Frailty, Suffering, and Vice: Flourishing in the Face of Human Limitations* (Washington, DC: American Psychological Association, 2017), 9.

[2]Cheryl Strayed, *Wild: From Lost to Found on the Pacific Crest Trail* (New York: Alfred A. Knopf, 2016), 189.

SUBJECT INDEX